ANIMAL

ATTRACTIONS

ANIMAL

ATTRACTIONS

NATURE
ON DISPLAY IN
AMERICAN ZOOS

Elizabeth Hanson

PRINCETON UNIVERSITY PRESS
PRINCETON AND OXFORD

Copyright ©2002 by Princeton University Press
Published by Princeton University Press, 41 William Street, Princeton, New Jersey 08540
In the United Kingdom: Princeton University Press, 3 Market Place, Woodstock,
Oxfordshire OX20 1SY

Library of Congress Cataloging-in-Publication Data

Hanson, Elizabeth, 1962–
    Animal attractions : nature on display in American zoos / Elizabeth Hanson.
        p. cm.
    Includes bibliographical references (p.).
    ISBN 0-691-05992-6 (alk. paper)
    1. Zoos—United States—History. I. Title.
QL76.5.U6 H36 2002
590'.7'373—dc21   2001055198

British Library Cataloging-in-Publication Data is available

This book has been composed in Palatino
Designed by Maria Lindenfeldar
Composed by Gary R. Beck

Printed on acid-free paper ∞

www.pup.princeton.edu

Printed in the United States of America

10 9 8 7 6 5 4 3 2 1

FOR BURKE

CONTENTS

ACKNOWLEDGMENTS

Writing a book requires the work of many people whose contributions take place behind the scenes. I am happy for the opportunity to acknowledge at least some of those contributions here. This study began as a Ph.D. dissertation at the University of Pennsylvania in the Department of History and Sociology of Science. I started thinking about the ideas presented here in seminars led by Robert E. Kohler and M. Susan Lindee. Their passion for ideas and history and their generous and intelligent guidance helped speed the completion of that project and sustain its transformation into a book manuscript.

Most of my research was carried out at the Smithsonian Institution Archives, with the support of graduate student and predoctoral fellowships. I am grateful to Pamela M. Henson, Division of Institutional History, for her enthusiasm and for introducing me to the community of scholars at the Smithsonian Institution. Navigating through the Smithsonian's extensive collections relevant to zoos was made easy thanks to the knowledge and good humor of the Smithsonian Institution Archives staff, in particular William Cox and William Deiss. Special thanks for research help are also due to Kay Kenyon, formerly the librarian at the National Zoological Park.

I am also grateful for the friendly assistance of Steven Johnson, archivist at the Wildlife Conservation Society, which holds the records of the Bronx Zoo. With his knowledge of the collections, the institution, and zoo history, he pointed me to many people and sources I would not otherwise have found. In addition, Linda Rohr provided helpful material on the history of the Franklin Park Zoo in Boston.

Over the years many colleagues and friends have read portions of the manuscript, and their advice and criticism shaped the book. Among them are Mark V. Barrow, Richard W. Bukhardt, Elizabeth Frank, Jeffrey Hyson, Katie Janssen, Vernon Kisling, Sally G. Kohlstedt, Henrika Kuklick, Libby Levison, Lynn Love, Susan Miller, Gregg Mitman, Don Moore, Lynn K. Nyhart, Nigel Rothfels, Helen Rozwadowski, David Schuyler, Stefanie Silverman, Elizabeth Toon, and Terence Young. My deepest thanks go to Susan D. Jones, who has been a friend, cheerleader, critic, and fellow student of human–animal relationships since the beginning of this project. Many others not named here have entertained long discussions about the minutiae of zoo history, sent me newspaper clippings about zoos and books about animal collectors, and provided general encouragement; I thank them too.

At a critical point in the completion of the manuscript Alice C. Lustig made it possible for me to take a leave of absence from my work at The Rockefeller University. Special thanks go to her and also my editors at Princeton University Press, Deborah Malmud and Thomas Lebien.

I am grateful for the moral support of my parents, Roger and Virginia Hanson, and their faith in the importance of books. My sister, Sandra Ferretti, created a first-aid kit for writer's block (it included lots of chocolate) that saw me through the final stretch. James and Margaret Anderson provided many kindnesses that made writing easier. I might never have embarked on this project, however, without Burke Anderson, whose keen eye for the natural world and respect for living things persuaded me to think critically about human relationships with nature. He also convinced me to forsake city life for the pleasures of the middle landscape, which looms large in the following pages. Burke's patience, encouragement, and sense of humor made it possible for me to write this book. It is for him.

ANIMAL

ATTRACTIONS

INTRODUCTION

On a rainy day in May 1988, a lowland gorilla named Willie B. stepped outdoors for the first time in twenty-seven years. Born in Africa in 1958, Willie B. had been captured by an animal collector and was delivered to the zoo in Atlanta, Georgia, in 1961, where he was housed by himself in an enclosure of concrete and heavy bars. Twenty years later, after complaints about the zoo's management, a television was provided to relieve his isolation. (He watched *M*A*S*H**, *60 Minutes*, and a save-the-zoo telethon.) Willie B. was listless and overweight, and hardly an ambassador for gorilla conservation, until the day when he tentatively looked out on the grass and trees of a new, naturalistic immersion exhibit at the renovated Zoo Atlanta. Other gorillas were released into the exhibit, and Willie B. soon adjusted to life in a social group, became a father, and evidently lived happily until his death in February 2000 at the age of forty-one. Willie B. had been the zoo's most popular attraction; a crowd of more than seven thousand people attended a memorial service in his honor. In his lifetime he had journeyed from being an object of voyeurism in a sterile cage to a muscular silverback, foraging for raisins and behaving like a gorilla. He achieved a kind of zoological fulfillment in his opportunity to live a more authentic gorilla life than he had behind bars, a transcendence in his return to nature.

Willie B.'s story parallels accounts—familiar through the news media—of how American zoos have introduced naturalistic exhibits in the last thirty years and begun to understand and implement ways of caring for animals so that they behave as they would in their wild habitats. During his life, zoos stopped collecting animals from the wild and started captive breeding programs. Americans changed the ways they wished to view animals in the zoo—bars became unacceptable. Field studies of gorillas, their behavior, and their natural environment—scientific knowledge unavailable to earlier generations—became widely available to the public and was applied to gorilla-keeping in zoos. And the importance of preserving gorilla populations and habitats, an inconceivable problem at the time Willie B. was captured and brought to Atlanta, emerged as the overriding educational message of gorilla exhibits at zoos.

Willie B.'s life also recapitulates the promise zoos have made to their human visitors for over a century. From their beginnings in the late nineteenth century, American zoos have offered people an escape from the cement, stress, and physical confinement of the city to a lush landscaped park. A trip to the zoo has long been presented as a journey into nature. And the idea that an excursion into the natural world is a healthy activity, restorative to mind and body and full of potential for self-improvement, has a long history. Part of the appeal of Willie B.'s story is that every zoo visitor can appreciate that the gorilla himself made the transition from life in a prisonlike cell to days spent lolling on a grassy hillside in the sunshine. In the late twentieth century zoo animals as well as zoo visitors have made an excursion into nature.

Of course, nature in the zoo presents all sorts of contradictions. What could be more unnatural than polar bears in Miami or giraffe in New York City? Zoos present a peculiar blend of nature and culture. They bring the natural world under the control of human civilization; they are parks that constitute a middle ground between the wilderness and the city, specially constructed meeting places for wild animals and urban Americans. This juxtaposition of wildness and civilization, naturalness and artificiality, makes up a large part of their fascination. And popular interest in zoos has been long lasting. Each year more than 130 million Americans visit zoos—more people than attend professional baseball, football, and hockey games combined.[1]

Most historical accounts of zoos look back to the animal collections of ancient history for precedents. Civilizations that accumulate wealth have long taken an interest in exotic animals. Queen Hatshepsut of Egypt sponsored expeditions to collect giraffe and cheetahs around 1400 B.C. Chinese emperor Wen Wang established a "garden of intelligence" before 1000 B.C. that included deer, antelope, and pheasants. In the fourth century B.C. Aristotle studied the animals sent back to Greece by Alexander the Great during his conquests. Exotic animals kept for pleasure, study, or as tokens of power retained their appeal in Europe during the middle ages on a smaller scale, the collections of the thirteenth-century Holy Roman Emperor Frederick II being the best known. During the Renaissance, explorers and traders collected live animals on their voyages, and royal menageries became symbols of status and power.[2]

Only a privileged few had access to such collections, however, and although they deserve study, they were the products of rather different historical circumstances than the zoos of the last century. Zoological gardens and parks for the amusement and education of the public are an invention of modern Western culture. In Europe, public zoos began to replace royal menageries in the late eighteenth century. Following the European example—in particular, imitating the London Zoo and various German zoos—Americans began building zoological parks in the late nineteenth century. The first zoo in the United States opened in Philadelphia in 1874, followed by the Cincinnati Zoo the next year. By the turn of the twentieth century, Chicago, San Francisco, Cleveland, Baltimore, Washington, D.C., Atlanta, Pittsburgh, St. Paul, Buffalo, Toledo, Denver, and New York City all had zoos. Tallies differ, but by all accounts, by 1940 there were zoos in more than one hundred American cities.[3]

The new zoos set themselves apart from menageries and traveling animal shows by stating their mission as education, the advancement of science, and in some cases conservation, in addition to entertainment. Zoos presented zoology for the nonspecialist, at a time when the intellectual distance between amateur naturalists and laboratory-oriented zoologists was increasing. Zoos also provided a new way for urban Americans to encounter the natural world, and they attracted wide audiences. By 1903, well over a million people toured the New York Zoological Society's Bronx Park each year, and in 1909 the Bronx Zoo's attendance was twice that of New York's more centrally located American Museum of Natural History. The Toledo, Ohio, zoo became a regional attraction. It reported a turnstile count of 1,216,400 in 1927—four times the city's population.[4]

Zoos quickly became emblems of civic pride, an amenity of every growing and forward-thinking municipality analogous to other institutions such as art museums, natural history museums, and botanical gardens. This study explores the cultural and physical landscape of the zoo rather than providing a chronological account of its institutional development, so a brief summary of the institutional story is called for here. Most American zoos were founded as divisions of public parks departments. They were dependent on municipal funds to operate, and they charged no admission fee. They tended to assemble as many different mammal and bird species as possible, along with a few reptiles, exhibiting one or two specimens of each, and they competed with each other to become the first to display rarities, like a rhinoceros. In the constant effort to attract the public to make return visits, certain types of display came in and out of fashion; for example, in the 1920s and 1930s dozens of zoos built "monkey islands." In the 1930s, the Works Progress Administration funded millions of dollars of construction at dozens of zoos. City zoos provided inexpensive recreation during the Depression and World War II. For the most part, collections were organized according to a loose taxonomic scheme—mammals, birds, reptiles—in a combination of houses and paddocks.

Although many histories of individual zoos describe the 1940s through the 1960s as a period of stagnation, and in some cases neglect, new zoos continued to open and old zoos changed their exhibits. In the 1940s the first children's zoos and farm-in-the-zoo exhibits were built. And after World War II an increasing number of zoos tried new ways of organizing their displays. In addition to the traditional approach of exhibiting like kinds together, zoo planners began putting animals in groups according to their continent of origin and designing exhibits showing animals of particular habitats, for example, polar, desert, or forest. By the late 1960s a few zoos arranged some displays according to animal behavior; the Bronx Zoo opened its World of Darkness exhibit of nocturnal animals. Paradoxically, at the same time as zoo displays began incorporating ideas about the ecological relationships between animals and their habitats, big cats and primates continued to be displayed in bathroomlike cages lined with tiles.

By the 1970s, a new wave of reform was stirring. Popular movements for environmentalism and animal welfare called attention to endangered species and to zoos that did not provide adequate care for their animals. Zoos began hiring full-time veterinarians and research scientists, and they stepped up captive breeding programs. Many zoos that had been supported entirely by municipal budgets began recruiting private funding and charging admission fees. In the prosperous

1980s and 1990s zoos built realistic "landscape immersion" exhibits, many of them around the theme of the tropical rainforest. Increasingly, conservation and the advancement of science moved to the forefront of zoo agendas, and educational programming expanded.

There is more to the story of American zoological parks, however, than a tale of progress and increasingly humane treatment of animals. The founding goals of entertainment, education, the advancement of science, and conservation sound surprisingly familiar today, but the meanings of zoos to both their audiences and their administrators have changed over the course of more than a century. The history of how zoo animals have been collected and displayed reveals a long-standing tension between nature appreciation as popular pastime and observing nature as scientific endeavor. As physical expressions of the uneasy pairing of wildness and civilization, science and popular culture, and education and entertainment, zoos have much to say about how Americans envision the natural world and the human place in it. This book seeks to understand how the zoo, an immensely popular and commonplace feature of American cities, took shape, and how relationships between urban people and wild animals have been constructed in the zoo landscape.[5]

American zoos came into existence during the transition of the United States from a rural and agricultural nation to an urban and industrial one. The population more than doubled between 1860 and 1900. And as more and more middle-class people lived in cities, they began seeking new relationships with the natural world as a place for recreation, self-improvement, and spiritual renewal. Cities established systems of public parks, and nature tourism—already popular—became even more fashionable with the establishment of national parks. Nature was thought to be good for people of all ages and classes: Fresh Air Funds for city children were established, as well as scouting, the Woodcraft Indians, and the Campfire Girls. Nature study was incorporated into school curricula, and natural history collecting became an increasingly popular pastime. As they hiked, camped, bicycled, and picnicked, Americans collected minerals, bird eggs and nests, plants, butterflies, shells, and birds and small animals to mount as taxidermied specimens. In addition, the first movements emerged to preserve nature and natural resources—to save the bison from extinction, for example, and to halt the hunting of birds for their decorative feathers.[6]

In addition, zoogoers at the turn of the twentieth century could learn about nature through popular essays and animal stories. Ernest Thompson Seton and Jack London wrote their best-selling books at this time. And it was in the realm of "realistic" stories about wildlife that a clash between science and sentiment in appreciation of the natural world was played out publicly, in the pages of *The New York Times* and elsewhere. Moral order in nature was an important theme of many stories, and writers also narrated from the perspective of animals or described the thoughts of wild animals. In 1903 John Burroughs, dean of American nature writers, launched an attack on the credibility of the writers of the new animal stories, later dubbed "nature fakers." Such writers, he argued, only masqueraded as naturalists; they sentimentalized and anthropomorphized the lives of wild animals, doing a disservice to people who wanted to learn the truth about nature. The issues played out in the nature fakers controversy also were evident among zoo audiences and zoo managers who anthropomorphized wild animals while seeking an educational experience at the zoo.[7]

At the same time, the fields of study subsumed under natural history in the nineteenth century were expanding, differentiating, and becoming professionalized into, among other things, taxonomy, experimental embryology, and genetics. Laboratory research gained prestige in the zoology departments of American universities. In general, the gap between professional and amateur scientific activities widened. Natural history had been open to amateurs and easily popularized. Laboratory research required access to microscopes and other equipment, as well as advanced education.[8]

While aiming for the cultural status of scientific institutions, and claiming a measure of truth in their representations of nature, zoological parks encouraged nature study and popular natural history. William T. Hornaday, first director of the Bronx Zoo, spoke out against teaching zoology in the laboratory as a method that "strives to set forth the anatomy of animals without adequately introducing the animals themselves." He advocated teaching children what he called practical zoology: "The pupil desires and needs to be taught about the birds of use and beauty, the big animals that are being so rapidly exterminated,

the injurious rodents, the rattlesnakes and moccasins, the festive alligator, the turtles." Forcing children to "write twelve paragraphs on the
mouth parts of a crayfish" would both kill their interest and deprive
them of "the immense amount of pleasure to be derived" from "a good
general knowledge of the most interesting animal species." The zoo
was a place to acquire this general knowledge.[9]

Although zoos were popular and proliferating institutions in the
United States at the turn of the twentieth century, historians have paid
little attention to them. Perhaps zoos have been ignored because they
were, and remain still, hybrid institutions, and as such they fall between the categories of analysis that historians often use. In addition,
their stated goals of recreation, education, the advancement of science, and conservation have often conflicted. Zoos occupy a middle
ground between science and showmanship, high culture and low, remote forests and the cement cityscape, and wild animals and urban
people. Furthermore, although zoos have always attracted diverse audiences, they are middle-class institutions. This may explain why historians of recreation and of popular culture, who have focused on
parks, for example, as arenas of working class rebellion, have over-
looked zoos. Zoos also may have been passed over by historians because of the lowly status of their animal inmates. The display of exotic
animals has been less interesting to scholars than the display of exotic
humans, which has figured in studies of ethnographic exhibits and
freak shows.[10]

Historians of science may have dismissed zoos as too entertaining,
connected to neither museum-based zoology nor laboratory science, or
simply unscientific "places of spectacle and dilettante scientific interest." To be sure, unlike European zoos, the first American zoos had
few ties to university zoology departments. The director of the National Zoo, when he visited the Amsterdam Zoo in 1929, commented—
without irony—that "It was interesting to find zoology being studied
in a zoo." The study of dead specimens in museums contributed far
more to the advancement of scientific knowledge around the turn of
the century than did observations of zoo animals. But amateur interest
in science bears examination both in itself and in its relationship to
professional science. This study has benefited from recent work that
focuses on how popular culture is made and used, that looks at issues

of scientific practice and the history of natural history, and that seeks to understand cultural representations of nature.[11]

The few scholars who have looked at zoos in their historical context have tended to focus on individual institutions and to emphasize the power relations implicit in the human gaze at caged animals, interpreting it as symbolic of imperial power over colonial subjects. Other writers have looked at zoo animals as stand-ins for humans, comparing zoos to prisons, for example, or analyzing the ways zoo visitors anthropomorphize animals. While zoos do express human power over the natural world, and until relatively recently they depended on colonial commerce to supply exotic animals, the process of collecting and exhibiting wildlife has been more complex than a display of dominance. Collecting, for example, has a history as a scientific endeavor, which zoos used in their attempts to raise their cultural status. It seems likely too that zoo audiences, particularly in countries without colonial empires, have seen zoo animals as more than surrogate colonials, and that the meaning of animals—elephants and eagles, for example—changes in different national contexts, and over time.[12]

Part of the impetus to analyze zoos as emblems of imperialism comes from their similarities to natural history museums. Museum scholars have looked to the ways in which museums ordered their collections, and at patterns of circulation through museums, for insights into relationships between knowledge and power, and into the means of social control exerted by bourgeois museum administrators over lower-class visitors. A parallel exercise could be performed with zoos. Early maps of zoos might reveal a narrative implied by the recommended order of viewing exhibits—a narrative of evolutionary progress, for example, reflecting the way some museums arranged their collections.[13]

But such an exercise is fraught with contradictions for both museums and zoos. Just as the availability of cheap natural lighting often dictated the placement of exhibits in museums, and helps account for their similarity to department store displays, the contours of the landscape played a role in the planning of zoos. An outcropping of rock might lend itself to a bear exhibit; a flat area could make a natural deer paddock. Clearly there was some order to the presentation of zoo collections, and it was often roughly taxonomic. But other considerations such as sanitation and ease of maintenance also played a role in determining the layout of zoos. Furthermore, order in the zoo was continually disrupted. A sick bird might temporarily be kept in the reptile house. Managers rearranged exhibits in order to attract visitors, and particularly beautiful or entertaining animals—flamingos, for example—might be placed near the zoo entrance, away from the rest of their kind. Few zoos maintained an internal unity over time that would allow the writing of a master narrative of order and power.[14]

Furthermore, such an approach favors the perspective of administrators—their ideas about the purposes of their institutions and how their plans were carried out. But zoo visitors experienced the displays in ways that managers did not anticipate, and they did not necessarily follow instructions. "Not for me is the admirable itinerary recommended in the guide-book," wrote one zoo lover, ". . . I make straight for the lions."[15]

Clearly zoos were planned in a way to distinguish them from earlier menageries, which were considered disorderly. But rather than interpreting zoos as examples of human dominance over nature, or emblems of imperialism at a time when the United States was gaining strength as a world power, this study situates them in the historical context of the American relationship to the middle landscape. It draws from environmental history, the history of natural history, and studies of popular culture to explore how zoos have used a curious and often uneasy blend of scientific research, education, and entertainment to negotiate their desire to create an authentic experience of nature for a popular audience. The following chapters explore the ways in which the layout of the zoo, the built form of specific exhibits, and the practices of collecting and displaying animals contributed to the definition of nature in the zoo.[16]

The development of American zoos has been powerfully influenced by their placement in large country parks planned at the turn of the twentieth century and by middle-class ideas about nature that formed in the United States in the nineteenth century. The creation of zoos as part of urban public park systems in the United States helps justify their consideration separately from European zoos, which had largely private origins. European models were adapted to American circumstances and values. Although European zoos appear occasionally in this study to provide points of comparison, they are not the focus here. Furthermore, this book does not provide an account of the institutional structure and management of particular zoos, or the details of their relationships to city governments and other cultural institutions. Rather, the aim here has been to understand, in broad terms, what the landscape of zoos, their displays, and the ways they have assembled their collections can tell us about relationships between city people and the natural world, and between science and popular culture. In addition, although zoo audiences have always been diverse and difficult to characterize, an effort has been made to examine their interaction with and contributions to the development of zoological parks. Zoos today often refer back to their founding goals: education, entertainment, the advancement of science, and conservation. The meanings of these goals have changed, but in the twenty-first century zoos continue to grapple with a problem that has remained consistent from their beginnings: how to convince their audience to appreciate wildlife.[17]

CHAPTER 1
ANIMALS IN THE LANDSCAPE

The entrance to the Philadelphia Zoo looks much the same today as it did in 1874, the year the zoo opened. Along the wrought iron fence where twenty-first-century visitors park their cars, zoogoers of the 1870s hitched their horses. Other nineteenth-century visitors walked from the nearest streetcar. Like visitors today, they crossed a broad plaza, perhaps meeting up with friends at the bronze lioness sculpture. Then they approached the twin Victorian gate houses that mark the entrance to the zoo. Recognizably the work of the Philadelphia architect Frank Furness, the gate houses attested to the cultural significance of the zoo. And with their fanciful brick facing and their gabled roofs, these romantic cottages beckoned visitors to the wonders within the park—a new kind of park—the first zoological park in the United States.

Unless they had traveled to zoos in Europe, visitors to the Philadelphia Zoological Garden in the 1870s anticipated a new experience. The exotic animals here were displayed in solidly built houses in a meticulously landscaped park. A zoo visitor might have seen a circus bear perform under a tent, but probably had not observed in one place a polar bear, a grizzly bear, brown, black, and cinnamon bears, and a "Borneo" sun bear, all inhabiting sections of a forbidding stone bear pit. Likewise, the rhinoceros, kangaroos, and giraffes at the zoo were as novel as their houses. In its variety and presentation, this was no common menagerie.[1]

FIGURE 1. Gate house at the Philadelphia Zoo in the late nineteenth century. Zoological Society of Philadelphia.

Yet the experience of going to the zoo would not have been altogether strange. A trip to the country to enjoy nature and to learn was a frequent and sought-after experience for middle-class Americans of the day. The Philadelphia Zoo was situated in a public park, one of many that were plotted and landscaped in American cities in the nineteenth century. The animal houses would have seemed familiar too; their design was reminiscent of fairground pavilions, or the retreats constructed at chautauquas.

The Philadelphia Zoo was more thoroughly planned and better financed than most of the hundreds of zoos that would open in the United States later. But in its landscape and its mission—to both educate and entertain—it embodied ideas about how to build a zoo that stayed consistent for decades and remain relevant today. The first American zoos were constructed in sections of larger public parks. Their origins are linked to the public parks movement at the turn of the twentieth century. In Philadelphia, the city leased land to the Zoological Society in a corner of Fairmount Park, a large tract on the outskirts of the city that had recently been annexed. A visit to the zoo was not only a chance to see animals, it was also an opportunity to experience the pleasures of a park, to escape the pressures of the city and admire natural scenery. An early visitor to the Philadelphia Zoo found the landscape as appealing as the animals: "the wealth of grand old trees, many of them gigantic in size, and the rolling surface of the land, render the whole place picturesque and beautiful."[2]

With the goal of public education, the Philadelphia Zoo attempted to set itself apart from other places where animals were displayed; many of the zoos that opened around the turn of the twentieth century also added "the advancement of science" to their agendas. It was important to state this because in Philadelphia, as elsewhere, the zoo was not the only animal show in town. Exotic animals had long provided entertainment in pubs and on street corners. Since the eighteenth century, an assortment of lions, camels, monkeys, and other beasts had toured the cities of the northeastern United States, along with human freaks and exhibits of curiosities. Popular museums, such as Peale's Museum in Philadelphia, frequently supported menageries, as did circuses. Philadelphia's zoo founders wanted to do something more refined. They were "extremely anxious to put the work of the institution as far as possible from the field occupied by the traveling menagerie." They explained, "We want to make of it an educator as well as a place of amusement." The atmosphere of a country park, and the reform goals of the parks movement, helped along the goal of the new zoos to instruct rather than merely to entertain.[3]

The civic leaders and naturalists who promoted zoos in American cities held a seriousness of purpose that corresponded with that of middle- and upper-class urbanites who campaigned for natural history museums "without humbug," as well as art museums, symphony orchestras, libraries, and other cultural institutions emblematic of progressive cities. Zoos became symbols of civic pride, yet they constantly had to shore up the claim to education that set them apart from mere menageries, this difference being "a distinction which the American popular mind is slow to recognize." Zoos used their landscape and layout to advance their mission. Like public parks, they provided a retreat for city dwellers, a balance of nature and culture. Other nineteenth-century landscapes that joined notions of nature, recreation, and enlightenment also provided models for zoos, for example, suburbs and college campuses. By building zoos to be reminiscent of these familiar settings, a middle-class ethos was enforced, which set zoos apart spatially and visually from popular entertainment, and above it socially.[4]

Collections of exotic animals have a long history preceding the zoological parks built in the United States, of course, and much has been written about them. American zoos took inspiration most directly from European zoos, and in some cases looked to them as specific models. In Europe, during the late eighteenth and through the nineteenth century, menageries that had once been the property of royalty increasingly became open to the public. Paris had the oldest public animal collection, founded as part of the Museum National d'Histoire Naturelle during the French Revolution. Police who cleared the streets of spectacles and other attractions deposited performing animals at the Museum in 1793. The survivors of the menagerie at Versailles—those not released or butchered during the revolution—joined the Museum collection the next year. In London, animals that had been displayed in the Tower menagerie, as well as animals collected in the colonies of the British Empire, were incorporated into a new zoo in Regents Park that opened in 1828. The collection was administered by the Zoological Society of London, which was granted a royal charter "for the advancement of zoology and animal physiology." Because the Paris and London zoos were founded to serve scientific research and education—they were intended as more than just exhibitions of curiosities—chroniclers of zoo history often point to them as the first modern zoos.[5]

By 1869, zoological gardens with varying claims to scientific status had opened in Dublin, Manchester, Bristol, Berlin, Frankfurt, Antwerp, Amsterdam, and Rotterdam. A spurt of zoo building took place in Germany in the two decades after 1858—zoos were founded then in Frankfurt am Main, Hamburg, Cologne, Dresden, Düsseldorf, Hannover, Karlsruhe, Münster, Breslau, Leipzig, and Stuttgart. In 1903 a guide book to European zoos was published; it described sixteen in German cities, four in Britain, and four in France. In most European cities, the zoo was managed and maintained by either a private zoological society or a joint stock company, and visitors paid a fee to enter the grounds. Animals were displayed in formally landscaped gardens, and the zoo was often situated close to the center of the city, or at least near a train station. European zoos featured bandstands and restaurants (in Berlin, the city's largest restaurant was in the zoo). They also maintained ties to scientific institutions through the appointments their directors held in university zoology departments.[6]

"[T]he great number, the size, the magnificence and the immense popularity of the zoological gardens" impressed Americans who toured Europe in the mid- and late-nineteenth century. Zoos played an important role in the civic culture of many European cities. In Amsterdam, for example, the zoological society, founded in 1838, served as a private social club for those committed to the development of a variety of cultural activities in the city. The society not only published journals of zoology, it also sponsored musical performances and commissioned new works for Amsterdam's leading orchestras. One American noted that "It is considered an honor as well as an advantage to belong to those zoological societies." Zoos were where "the aristocracies of intelligence, of wealth, and of birth" met to stroll and to socialize. Striving to emulate this bourgeois culture, some Americans led campaigns for zoological parks at home. The organizers of the first American zoos, in Philadelphia and Cincinnati, looked to European zoos for guidance.[7]

William Camac, a Philadelphian trained as a physician and active in many scientific and social organizations, had visited Europe in the 1850s. When he returned he organized an elite group of local naturalists and distinguished citizens to form the Zoological Society of Philadelphia, taking the London Zoo as a model for organization. Camac pursued recognition of the Society from the state of Pennsylvania, and in 1859 the state legislature chartered the Society to oversee "the purchase and collection of living wild and other animals, for the purpose of public exhibition" and "for the instruction and recreation of the people." The Civil War distracted attention from plans for the zoo—the Society "languished and lay dormant" during the war, in Camac's words. But under Camac's leadership it was revived with renewed enthusiasm in the 1870s, still looking to London for guidance. As construction on zoo buildings was beginning in Philadelphia's Fairmount Park in 1873, the Society sent its engineer to study the animal houses in the London Zoo. An experienced animal collector was hired to stock the zoo and then serve as head animal keeper. When the Philadelphia Zoo opened to the public in 1874, it was the first animal exhibition in the country to be housed in permanent buildings and tended by a full-time staff. And according to one well-traveled writer, within just a few years of opening it achieved the high standard it had aimed for: "the air and general appearance of famous long-established like institutions in Europe."[8]

In the meantime, a prosperous community of German immigrants in Cincinnati, remembering the zoos in Germany as important cultural institutions, sought to build a zoo in their adopted home. The German-born Andrew Erkenbrecher, animal lover and owner of a starch manufacturing company, led the effort to bring together a group of prominent citizens—naturalists, businessmen, and civic leaders—to form a zoological society, which was established in 1873. The new society was organized as a joint stock company, and its stated purpose was typical of new zoos: "the study and dissemination of a knowledge of the nature and habits of the creatures of the animal kingdom." Members of the zoological society also expected the garden to be "a profit to the stockholders, a credit to the city, and a continual source of improvement to its visitors." The society patterned its zoo after the one in Frankfurt and sent a representative to Germany to recruit its first superintendent. The zoo hired Dr. H. Dorner in 1875, luring him away from his post as scientific secretary of the Hamburg Zoological Garden. In the next decades zoos were founded in other American cities as well, with various combinations of private and public support.[9]

Zoos and the Movement for Public Parks

European zoos continued to be sources of instruction and inspiration to the directors of new American zoos. At every opportunity they toured European zoos to seek advice on techniques of animal care, admire new zoo buildings, and buy animals. But imitation of European zoos was not the only impetus for building zoos in the United States, and probably not the most important one. American zoos were also products of the movement to create public parks. Late-nineteenth-century anxiety about the detrimental effects of the city on both health and morality led to the establishment of large country parks on the outskirts of many American cities. Middle-class reformers advocated parks as helping to contain the threat posed by urbanization to moral and social order. American city planners created parks as pieces of country in the city, restorative retreats that would offset the stress, noise, grime, overstimulation, debauchery, and disorder of city life. They built parks in reaction to urbanization and industrialization rather than in imitation of European parks. In the same period, similar sentiments guided the formation of national parks where city dwellers could seek healthful recreation in nature.[10]

The appeal of such middle landscapes—gardens, parks, or other natural landscapes situated outside the overstimulating city but short of the primitive wilderness—has a long history in American culture and in Western thought. Such places integrate nature and culture, joining pastoral scenery and civilization. They are the landscapes of Arcadia, of Thomas Jefferson's garden, and of American writers like Thoreau. They represent a moral landscape—a landscape of virtue—as well as a place in nature, "a symbolic repository of economic, political, religious, and aesthetic values such as concern for self-education, the value of the arts, the self-enrichment of recreation, the moral precepts of Protestant Christianity, and the impact of physical environment on human behavior."[11]

With urbanization and industrialization in the nineteenth century, the middle landscape appealed increasingly to middle-class Americans. They sought refuge in nature, especially in their leisure time. Seeking aesthetic and spiritual fulfillment, businessmen, clerks, preachers, and professors went hiking and camping. They vacationed in the mountains and by the seashore, and they pursued woodcraft and nature study. The White Mountains, Niagara Falls, and scenic destinations farther west became tourist attractions as the century progressed. The development of railroads allowed more people access to rural scenery—"Routes are laid out to beautiful lakes that had been visited only by occasional fishermen or hunters, or to picturesque spots on river or sea-shore"—and even promoted its appreciation in the view of one writer: "The love of natural scenery—the most universal of the aesthetic passions—finds a means for its gratification and cultivation in the rural trolley line." With trolley transportation middle-class people also began living in suburbs and commuting to work in the city. In going "back to nature" through such activities Americans created new landscapes, including parks, as well as religious camp meeting sites, communitarian settlements, chautauqua sites, college campuses, and suburbs. Such landscapes combined recreation, enlightenment, and a commitment to the benefits of the outdoors. Zoological parks were among these new landscapes.[12]

Frederick Law Olmsted is probably the best known of nineteenth-century park planners, and the founder of the profession of landscape architecture. His design, with Calvert Vaux, of Central Park in New York City in the 1850s influenced the construction of public parks for decades afterward. Park planners across the country consulted Olmsted, and he also designed other landscapes that were parklike in both substance and spirit—the grounds of private estates like Biltmore, university and land-grant college campuses, and one of the first planned suburbs, Riverside, built outside Chicago. Olmsted believed that meditating on nature in the surroundings of a large country park would offer psychic restoration to tired city workers. The exercise of walking and the appreciation of scenery would lead to mental and moral uplift. The planners of so-called pleasure ground parks across the country incorporated Olmsted's ideas in their designs.[13]

Integral to the design of pleasure ground parks was the idea derived from English garden theory that parks should be informal: stands of native species of trees should be planted, and flower beds and clipped shrubs avoided. But the beauty of natural landscapes needed to be heightened or enhanced. Nature on its own—a city-owned tract of undeveloped land, for example—was not automatically a park. In some cases, an appropriate natural-looking landscape had to be constructed from scratch. Chicago's Lincoln Park, for example, had been "covered by sand dunes, upon which there was no vegetation, and by patches of scrub oaks and a wilderness of tangled vines and weeds." Parkgoers in the late nineteenth century were aware that the "great beauty of Lincoln Park is in no wise due to original gifts of nature," and they enjoyed the deception. A guide book described a scene in Lincoln Park near the zoo's bear pits: "[T]he art of the landscape gardener is wholly disguised by the wealth of foliage everywhere covering the marks of his handiwork. Even this little stairway, made of rough stones, might have been carved by nature herself, whose skill in stone-cutting and molding, through the action of water, is scarcely less surprising at

FIGURE 2. The "Jungle Walk" at the Bronx Zoo in 1907: a middle-class middle landscape. © Wildlife Conservation Society, headquartered at the Bronx Zoo.

times than that of man's guided hand." Park planners, following Olmsted's models, created winding paths that gave the visitors a succession of views, in contrast to the grid of city streets, and they avoided laying paths or planting trees in straight lines. No statues or other explicit evidence of human artifice were permitted. Open lawns provided relief from the confining buildings that lined city streets. Visitors found parks like Lincoln Park "full of picturesque spots . . . [that] meet one at every turn, and always unexpectedly, thereby intensifying the charms of the Park and making a visit there a continued succession of surprises." Zoos added a variation to this theme by placing animals in the pastoral landscape. Landscape and animals complemented each other at the Bronx Zoo, where, "In October, when the splendid groves of beech, oak, and maple along the eastern ridge put on all the glorious tints of autumn, . . . then are the Elk also at their best."[14]

What a park should look like may have been generally agreed upon, but how a park should be used was the subject of much debate. Beyond quieting the overstimulated minds of urban professionals, reform-minded park planners intended parks to serve as a means of uplift for the working class. They associated appreciation of nature with self-improvement and unstructured outdoor recreation with both good health and good morals. They advocated "enjoyment"—a subtle compromise between superficial amusement and serious study of art or nature—in their parks. For Olmsted, the stress of city life included the stress induced by tension between social classes. Well-designed pastoral urban parks, in his view, helped to ensure social tranquility and order.[15]

To promote these goals, park planners banished reminders of the city such as advertising and commercial activities. They also differentiated public parks from the privately run picnic grounds and gardens that were precursors to amusement parks. Most cities had a place like Woodward's Gardens in San Francisco, Longfellow Gardens in Minneapolis, Elitch's in Denver, or the Elysian Fields in Hoboken, where in addition to strolling and admiring the flower beds, visitors could enjoy drinking, dancing, and popular theater, and sometimes view a menagerie. Such parks were eclectic in design, and interpretation of nature there ran to topiary, labyrinths, and beds of flowers planted in geometric designs. Novelties like exhibits of panoramas and equestrian acts drew crowds. Trolley car companies developed similar parks in the suburbs of Augusta, Akron, Boston, and elsewhere. These parks were meant to encourage people to ride the train, and to stimulate land sales. They were also popular. According to one writer, "Thousands . . . now are enabled . . . to enjoy . . . a visit to one of the great recreation-grounds run by the street-railway company, with all sorts of attractions—band concerts, variety performances, a menagerie, swings, teeter-boards, roller-coasters, fireworks, etc."[16]

Municipal park planners, however, banned alcohol and amusements such as polka music (because it was associated with dancing). Restaurants also verged on decadence, in their opinion, and they even debated the merits of refreshment as innocuous as water fountains. Among these and other issues, park planners also deliberated over whether attractions like zoos, botanical gardens, meteorological ob-

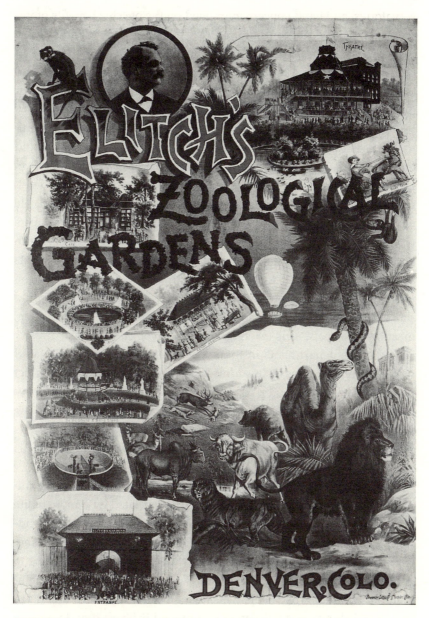

FIGURE 3. Elitch's Gardens in Denver had a zoo among attractions such as hot air balloon rides and a re-creation of the sinking of the *Titanic*. Denver Public Library, Western History Department.

servatories, and band concerts impinged on the psychic benefits of immersion in nature, and thus ran contrary to the purpose of parks. In the view of landscape architect Andrew Jackson Downing, whose work Olmsted admired, parks should be omnibus cultural institutions, combining the recreational uses of a pleasure garden with institutions for public education such as botanical gardens and zoos.[17]

Olmsted himself liked zoos, but was opposed to too many large animal houses consuming precious green space. A display of deer could enhance the rural scenery, but a zoo like the one in Regent's Park in London distracted from the regenerative power of the natural landscape. The London Zoo seemed particularly to offend Olmsted. He wrote that it interfered with "the purpose of providing great numbers of people living in a compactly built town all around such a place, with an opportunity to get quickly out of the scenery of buildings, streets and yards into scenery to be formed with a view to supplying a refreshing contrast with it." In Olmsted's view, Regent's Park "would much better serve such a purpose if it had as much more interior, open, meadow-like space as is taken up by the Zoological Garden."[18]

City parks departments in the United States, however, were faced with hundreds of acres of "meadow-like space." They needed to attract people to their parks, and park users wanted entertainment. For many urbanites "the loveliness of the landscape" alone was "not an adequate lure" to the park. As Olmsted's successors in his firm conceded, "It is probable that for most of the population the enjoyment of scenery is in a more or less embryonic state."[19]

Zoos may have been allowed in parks partly because looking at animals was considered a more moral activity, associated with more polite standards of behavior, than, for example, going to the theater. Mid-nineteenth-century circuses, which had previously toured separately from animal shows, added menageries to their troupes in an effort to entice a puritanical audience otherwise opposed to such entertainment. In their advertisements they sometimes attempted to legitimize the animal shows as a "dignified and refined Sabbath day diversion" by making reference to scripture. They billed the hippopotamus as "the blood-sweating behemoth of holy writ, spoken of in the book of Job," and painted lion cages with scenes of the story of Daniel. Showmen like P. T. Barnum catered to respectable audiences using different tactics, in part by attracting the interest of reputable naturalists with live animal exhibits.[20]

The fact that European zoos attracted a bourgeois audience also helped persuade park planners to build zoos. But even though the directors of these new zoos were impressed by both the high status of European zoos in civic culture and the scope of their animal collections, they did not transplant European zoological garden design to the United States unchanged. Americans found European zoos cluttered. Most European zoos occupied thirty acres of land or less, often near the center of the city. The Amsterdam Zoo, on twenty-five acres, seemed "a little shut in by the houses that surround it," and its lawns were not "sufficiently extensive," according to one observer. The London Zoo crowded 2,369 specimens into sixty-four houses, aviaries, sheds, and ponds on its thirty-one-acre grounds, according to an 1896 tally. To an American visitor, this constituted "a bewildering array of living creatures."[21]

Furthermore, American visitors objected to the architecture of European zoo buildings. Expensive and elaborately ornamented, they were out of keeping with both the modest means of city park departments and the aesthetic of rural pleasure grounds. In European zoos "colonial style" architecture, celebrating imperial conquest, was the fashion. Thus elephants were housed in a reproduction of a Hindu temple, and ostriches in a structure adorned with Egyptian hieroglyphs. One visitor to the Antwerp Zoo remarked that "Each large building for animals is called a 'palace,' and not one belies its name." But for American visitors such artificial surroundings distracted from the experience of observing nature. Another writer dismissed these "rather pretentious buildings which fail to present the animals in a really effective manner. Perhaps with the motive of emphasizing the excellence of the picture the frame has been so elaborated and magnified as to make the picture seem a mere incident." Palatial buildings also could be difficult to maintain. The elephant house in Berlin—a building famous throughout Europe—was "suffering from the weather and is difficult to keep in repair." Elaborate buildings set close together gave European zoos the atmosphere of a fairground, or worse: "[I]n Berlin," wrote a visitor from the Smithsonian Institution, "fantastic designs have been adopted, and some of the buildings are painted on the outside in gaudy colors, rivaling the wagons in a circus procession."[22]

American zoo planners faced different circumstances from their European counterparts. They had more park land to work with and in most cases lower budgets. Land annexations made in the nineteenth century gave cities access to large tracts to incorporate into their park systems. Zoo planners had access to parks encompassing hundreds or thousands of acres—Swope Park in Kansas City, Druid Hill Park in Baltimore, Roger Williams Park in Providence, Forest Park in St. Louis, and Como Park in St. Paul, among others. In any case, whenever possible, they planned zoos in spacious surroundings where the human encounter with nature would be restorative rather than overstimulating, in keeping with ideas about park design in general. Even the Philadelphia and Cincinnati zoos, which were designed most closely on European models, were situated outside the city in areas long regarded as picturesque. The Philadelphia Zoo, although relatively small, enclosed within its borders the house known as Solitude, built in 1785 by John Penn, grandson of William Penn. The estate had been a landscaped refuge from Philadelphia for nearly a century before the Zoological Society acquired it. Cincinnati Zoo founders leased a hilly, wooded site three miles from downtown, a logical choice for attracting day-trippers from the city. Since the early nineteenth century the area had drawn people from Cincinnati, first to religious revivals held in the heights, and then as vacationers; by the 1850s the area served as the setting for borderland life in popular fiction. When the zoo was built in the 1870s, the surrounding land had become a suburb with a reputation for its scenery, removed from the city but within commuting distance.[23]

With more land to work with, and a naturalistic aesthetic, zoo planners conceived of their parks in different terms from the formal urban gardens that were European zoos. American zoos would be places where at least some animals would be exhibited in natural surroundings. Among zoo proponents, William Temple Hornaday, the naturalist, collector, taxidermist, and conservationist, was most articulate about the idea of zoos as spacious animal parks. Hornaday described his ideal zoological park as "midway between the typical 30-acre zoological garden of London, Paris, Antwerp, or Philadelphia, and the

FIGURE 4. The ostrich house at the Berlin Zoo was built to resemble an Egyptian temple. In *Streifzuge durch den Zoologischen Garten Berlin mit der Zeiss Ikon-Camera,* 1927, Verlag des Actien-Vereins des Zoologischen Gartens zu Berlin.

great private game reserve." It was a place "in which living creatures can be kept . . . in spaces so extensive that with many species the sense of confinement is either lost or greatly diminished, yet at the same time sufficiently limited that the animals are not inaccessible or invisible to the visitor."[24]

Hornaday arrived at this idea from a concern for the conservation of American big game. He was chief taxidermist at the Smithsonian Institution's National Museum in the 1880s, when he determined to mount an exhibit of bison. For Hornaday, a lifelong collector of dead specimens, the events that followed amounted to a conversion experience. By the 1880s, the bison, long declining in numbers, had been nearly exterminated by hunters collecting trophy heads and robes, or just shooting for sport. During two trips to Montana in 1886 Hornaday was able to collect enough animals for his mounted group. But his difficulty in securing the specimens set him on a campaign to save the bison; Hornaday had been transformed into a conservationist. When he returned to Washington, D.C., he began campaigning for a zoological park with large enough grounds to serve as a breeding facility for bison and other endangered animals.[25]

Hornaday's first step toward this goal was to create a department of living animals at the Smithsonian. Bison, as well as bears, deer, foxes, prairie dogs, and a golden eagle, among other animals, were housed on the Washington Mall near the Smithsonian "castle." Hornaday lobbied his superiors for a more permanent zoo, and in 1889 he was charged with overseeing the surveying of a site in Washington, D.C.,'s Rock Creek Park, which was "situated at a convenient distance from the city in a region of remarkable natural beauty." The Smithsonian Institution took possession of 166 acres of Rock Creek Park in 1890 as the site of the National Zoological Park. Frederick Law Olmsted was called in, along with his stepson John C. Olmsted, to prepare a plan for the new zoo. Soon after, however, Hornaday had a falling out with the Secretary of the Smithsonian, Samuel Pierpont Langley, over his title and duties at the new zoo. Langley was unwilling to delegate responsibility to Hornaday, and Hornaday left Washington without seeing the zoo completed.[26]

Five years later, however, Hornaday again had the opportunity to carry out his vision of a zoological park. In 1896 the newly formed

FIGURE 5. Bison behind the Smithsonian Institution "castle" on the Washington, D.C., mall, circa 1887–1890. Photo Archives, National Zoological Park, Smithsonian Institution.

New York Zoological Society invited him to help plan a new zoological park and become its first superintendent. Members of the New York Zoological Society held a similar philosophy to Hornaday with respect to zoo planning. The group was made up of members of the Boone and Crockett Club, wealthy sportsmen-naturalists, including Theodore Roosevelt. They wanted to build a zoo on even larger grounds than the National Zoological Park, with space for species of American big game to roam and to breed. In addition to establishing a public zoo, the Society's goals were "the preservation of our native animals" and "the promotion of zoology." Hornaday agreed to join the project, and was put in charge of choosing a zoo site. After considering land in Crotona, Pelham Bay, and Van Cortlandt Parks, Hornaday decided on Bronx Park as the best location. His description of it harkens back to the literary mode of the garden as a place of untouched splendor, bounteous, and full of potential that could be realized with the appropriate human intervention. Hornaday wrote: "In the southern portion of Bronx Park we find a wonderful combination of hill and hollow, of high ridge and deep valley, of stream and pond, rolling meadow, rocky ledge and virgin forest of the finest description . . . here it remains only to skilfully [sic], artistically and sensibly *adapt* the work of nature."[27]

Hornaday was also practical in his "absolute requirements in an ideal zoological garden." Accessibility by public transportation ranked high on his list, and the Bronx Park met this criterion as well. Hornaday himself timed the trip by rail from City Hall: it took an hour and ten minutes and cost ten cents one way. In other cities, too, zoological parks built on the edge of the city were within a reasonable commute. Boston's zoo opened in 1913 in Franklin Park, the largest of Olmsted's Boston parks, and the one farthest from downtown. Columbus, Ohio, built its zoo in the 1930s sixteen miles north of the city, at the end of a scenic drive along the Scioto River and a "municipal system of parks and picnic areas that extends almost the whole distance between the zoo and the city." Likewise, eighty-three acres of land donated to the Chicago Plan Commission in 1919 by Edith Rockefeller McCormick for the construction of a new zoo were located in the borderlands fourteen miles west of the city, "accessible by fast interurban train service, elevated and surface street cars, and by automobile, of course." Hornaday was consulted on the property for the Chicago zoo. Although he found the site rather flat, he praised it: "The land is in the right spot."[28]

Park Reform and Menagerie Reform

In adopting the landscape of the pleasure ground park, zoo proponents like Hornaday were also engaged in the social reform aspects of park planning. (Hornaday was more explicitly a reformer than most zoo directors—he had been active in the temperance movement in the 1880s.) Like the pleasure ground parks in which they were situated, zoos embodied an elite ideal of recreation. One role of a zoo in a public park was to further goals of educating working class and immigrant visitors to middle-class standards of behavior. As one journalist put it, "It matters little whether Michael Flynn knows the difference between the caribou and the red deer. . . . It does matter a lot, however, that he has not sat around the flat disconsolate, or in the back room of the saloon, but has taken the little Flynns and Madam Flynn out into the fresh air and sunshine." In New York, Hornaday carried out a celebrated "rubbish war," posting throughout the zoo "150 cloth signs, printed in English, Yiddish, Italian, and German, forbidding the scattering of rubbish, and directing that it be placed in the baskets," under penalty of arrest. The zoo enlisted the police to arrest "foreigners, especially Italian laborers" who shot songbirds on the zoo grounds for food. Like other American parks, zoos did not serve alcoholic drinks at their refreshment stands, and the Bronx Zoo battled constantly to keep them out: "The demand for new entrances at various points, sometimes merely for the convenience of neighboring saloons, has been and still is continuous," wrote the Society's secretary in 1910. Social reform, whether by bringing middle-class city dwellers to a rural setting, or by bringing civilization to the working class in the form of public parks, was part of the agenda of the parks movement generally.[29]

Zoos, however, undertook an additional kind of reform. The new zoos reformed ideas about the meaning of wildlife by taking exotic animals out of nature and other settings and placing them in the middle landscape of a park. In what could be called menagerie reform, zoos provided an alternative to viewing animals in a haphazard array of menageries and sideshows. They also culled animals from inaccessible game parks and from wilder, far-flung jungles, deserts, and arctic icebergs. Putting these animals in zoos stripped them of their other habitats—both the circus ring and their places in ecology. Zoos reformed the image of animals by placing wildlife in a landscape of visual and behavioral conventions that were familiar to middle-class viewers.

For Hornaday, reform began with the word "zoo" itself. To him it represented the menageries and circuses he was trying to distance himself from. In the nineteenth century, the "zoo" had become the nickname for the London Zoological Society's animal collection in Regent's Park, after a popular dance hall song with the lyrics, "The OK thing to do, on a Sunday afternoon, is to toddle in the zoo." But in Hornaday's opinion the word zoo connoted something sordid. As Hornaday judged the term: "a 'zoo' is a small, cheap and usually smelly affair, with little scientific standing, or none at all." In contrast he preferred to call the kind of animal collection he advocated "a zoological park, or zoological garden," which denoted "a public institution of large size and more or less dignity." He considered the "Bronx Zoo" to be an "odious nickname" for his own New York Zoological Park.[30] Hornaday's bluster, of course, did little to deter anyone from calling his park the Bronx Zoo, and his self-righteousness was not lost on the local New York newspapers. The *Press*, for example, poked fun at him with the following doggerel:

> My name is William Hornaday—
> A trifle pedagogical.
> DIRECTOR I! With ALL to say!
> My Park is Zoological.[31]

In any case, Hornaday's distinction between zoos and menageries was more than a divide between learned and lowbrow. It was also a geographical line, a difference in landscape. Hornaday's ideal for zoos was to display the animal in a park, with extensive paddocks; this was in contrast to dime museums, pet shops, and private menageries, which exhibited animals in cages stacked on top of one another. Menageries, by definition, were crowded and disorderly; they displayed the same threats to social order as did immigrants housed in city tenements. Zoological parks, as Hornaday preferred to call them, belonged to the middle landscape.

Not all zoos had beginnings as well organized and financially sound as the Bronx Zoo. In the new city parks, animal collections accumulated unbidden. Such "menagerie slums," as Hornaday referred to them, often began as gifts of a few deer. In 1897, for example, the park boards of St. Paul and Minneapolis donated a total of three deer to fashionable Como Park. The tremendously popular menagerie in New York City's Central Park had grown in a typically haphazard way starting in the 1860s, to the dismay of Frederick Law Olmsted, who described its origins:

> [A] few small animals were received, most of them, having been, I think, pets of children who had died, or who were leaving town. As additional animals were, from time to time, presented, the Commissioners, never liking to decline gifts to the city, had pens and sheds prepared for some. . . . At a later period, the Director was allowed to make the collection more interesting by ex-

changes; and, still, later, by purchases; and so by successive, un-premeditated steps, the present conditions have been gradually approached.

Namely, the menagerie was "ill-arranged, ill equipped, not adapted to economical maintenance." The working-class people who crowded the free exhibits on Sundays also violated bourgeois decorum, and when P. T. Barnum housed his circus animals in the Central Park menagerie, it came uncomfortably close to commercial entertainment. The private New York Zoological Society represented a more genteel approach to zoo planning; in the 1890s, some Central Park commissioners, as well as some property owners opposite the old menagerie, hoped that the "poor man's monkeys" would be transferred to the new zoo in the Bronx.[32]

Most zoos in the United States had beginnings like the one in Central Park rather than the one in the Bronx. When, as in Denver, a gift of a bear was made to the mayor, or, as in Baltimore, several swans were donated to the city park commissioners, these animals were put in a public park. In other cities, like St. Louis and Buffalo, animals left behind after world's fairs added to an already existing city park menagerie. Bankrupt circuses provided another supply of animals to city parks. In Atlanta, local businessmen purchased the animals from G. W. Hall's Circus and Bingley's English Menagerie when the show ran out of money during an 1889 tour and donated them to the city. Detroit's first zoo was also the result of a circus going bankrupt. Such unplanned exhibits failed to fulfill the proper functions of a zoological park—the decorous combination of education and entertainment. In short order, these animal collections became too big to ignore, and parks departments planned the construction of zoos and acquired more select groups of animals. Private citizens organized into zoological societies often came to the aid of park officials in bringing order to these haphazard collections. They built animal houses and provided full-time keepers. Running a zoo gradually became a more professional affair. By the late 1910s, directors of city parks departments debated the purpose of zoos and exchanged tips on how to build and maintain them in the pages of *Parks & Recreation*, the official organ of the American Institute of Park Executives. In 1924, the first professional association for zoo managers was formed from this group, the American Association of Zoological Parks and Aquariums (AAZPA). Menagerie reform had achieved national recognition.[33]

Planned Communities

Although situated in parks, and planned with similar landscapes and expectations for visitor behavior, zoos were more than parks. Exhibiting animals required buildings, and an arrangement of such structures was necessarily more formal than the meadows, thoughtfully arranged clumps of trees, and placid lakes of the typical country park. A zoo was in effect a planned community, albeit with animal inhabitants instead of humans. It required housing, facilities for food preparation and for waste disposal, administration buildings, services for visitors, a defined perimeter, and points of access, usually gates of some sort. In architecture and layout, zoos looked much like other planned communities of the turn of the century that were built in parklike settings and incorporated middle-class ideas about nature. For human visitors, zoos combined recreation, enlightenment, and a commitment to the benefits of the outdoors, and thus were similar to other "middle landscapes of the middle class," such as suburbs and campuses, as well as chautauquas and religious camp meeting sites. In the arrangement of their buildings, zoos were spatial analogs of such communities.[34]

Early zoo plans bear a remarkable resemblance to the plans of nineteenth-century suburbs, which were sylvan retreats for those who could afford both an expensive home and train fare to jobs in the city. The first of these communities was Llewellyn Park, constructed in the 1850s in West Orange, New Jersey—just twelve miles from Manhattan by train. Instead of the gridiron pattern of streets common in towns, architect Alexander Jackson Davis, the planner of Llewellyn Park, laid out winding roads that conformed to the contours of the landscape. Davis allowed for open space—a fifty-acre "ramble"—to preserve the natural character of the development. Single-family homes were built on large lots in Gothic, Swiss chalet, and Italianate styles.[35]

The grounds of the Philadelphia Zoo in the 1870s reflected a similar romantic sense of planning. Like Llewellyn Park, the Philadelphia Zoo was a gated community in a wooded landscape. The original animal houses, described as "quaint and picturesque" by a contemporary critic, were not too far removed in design from the fanciful cottages of Llewellyn Park, if less elaborate. With its gates, lawns, woods, and houses, the zoo looked like a rural retreat for the middle class. Early images of the zoo reinforced this interpretation. A typical one, published in *Harper's Magazine*, shows a panoramic "view from the heights," with the zoo in the foreground, a style of imagery that celebrated the beauty of the crafted landscape on the outskirts of the city.[36]

That zoo plans resembled suburbs was no accident; zoo architects often had hands-on experience with building both planned communities and individual country houses and estates. Cincinnati's zoo, for example, was constructed under the supervision of Park Superintendent Adolph Strauch, who had designed estate landscapes in the suburb of Clifton. The National Zoo hired Boston architect William Ralph Emerson, known for his New England shingle country houses. Arthur Shurtleff, who had worked a dozen years in Olmsted's firm, created the master plan for the Franklin Park Zoo in Boston. And although Frederick Law Olmsted gradually retired from practice in the 1890s, his sons carried on his firm's work, which included advising the planners of the National Zoological Park, the Denver Zoo, and the Audubon Park Zoo in New Orleans.[37]

Not all zoos—or suburbs—followed the same plan, of course. Many zoos, like some planned communities, were designed with a formal town center surrounded by farms and fields. They arranged substantial stone houses for lions, birds, and reptiles around a pedestrian mall, or a rectangular green, like a village square, with a sea lion pool in place of a fountain. In the "downtown" of Chicago's Brookfield Zoo, a 100-foot-high goat mountain was constructed that would "be a landmark for miles around," like a church steeple. Winding paths branched away from this central focal point to the paddocks of deer and goats, on the rural outskirts of the animal community. Chicagoans referred to the Lincoln Park Zoo as "animal town." In Boston's Franklin Park, "heavy plantings of high woodland" shielded the intrusion of the zoo-village from the sight of other park users to reinforce the impression of being in the country. Other zoos lacked a clearly marked "downtown," yet were still suburban in design. Zoos like the National Zoo, which was too hilly for a central square, were laid out entirely as rustic pathways, with large single-family animal houses on spacious grounds.[38]

In addition to the suburb, the university campus provided another model for planned communities in the nineteenth century. Maps of many zoos might equally well be mistaken for university campuses, with zoo administration buildings, cafeterias, and the various animal shelters standing in for libraries, lecture halls, and fraternity houses. Campuses were characteristically located in the countryside, following Thomas Jefferson's notion of the self-contained "academical village" removed from the distractions of city life. Jefferson's design for the University of Virginia was widely imitated at new campuses throughout the nineteenth century. He arranged professors' pavilions and student dormitories around an open square, with a domed structure at one end to provide a visual focal point. It was a scheme that also worked well in zoo design. The Hermann Park Zoo in Houston followed such a U-shaped plan, as did the Washington Park Zoo in Milwaukee.[39]

Circuses, Amusement Parks, and Fairs

The resemblance of zoos to landscapes like suburbs and university campuses comes into clear focus when zoos are compared to other places where exotic animals were displayed. Many animal displays remained outside the fold of the new zoos. Elephants and camels, lion tamers and organ grinders, continued to perform at circuses and fairs and in vaudeville. In fact, despite the efforts of zoos to distinguish themselves from circuses, the two held much in common. Personnel frequently crossed over. At the turn of the century, a circus animal trainer who tired of life on the road could apply for zookeeper work. In 1891, for example, the National Zoo hired William Blackburne away from the Barnum and Bailey circus; he remained head keeper there until his retirement in 1944 at the age of 87. Circus man John Robinson was among the founders of the Cincinnati Zoo. And soon after its founding the Cincinnati Zoo hired Sol Stephan, a former elephant keeper, as director.[40]

FIGURE 6. Orangutans exhibit their table manners as they take tea at the Bronx Zoo. © Wildlife Conservation Society, headquartered at the Bronx Zoo.

Animals too could belong to both the circus and the zoo. The National Zoo's director from the 1930s through the 1950s was a long-time circus fan and expanded the zoo collection by allowing circus animals to spend winter, the off season, there. Other celebrity animals boarded at zoos as well. Clyde Beatty, the animal trainer and film star, loaned his lion and tiger to the Detroit Zoo in 1946. Some circus animals became permanent additions. The National Zoo's first elephants, for example, were gifts of the Adam Forepaugh shows. And zoos could serve as retirement homes for elderly animal performers, like Velox, a polar bear in the Denver Zoo. Zoos and circuses also relied on the same animal dealers to supply their stock, and sometimes traded with each other.[41]

Zoos and circuses also both featured performing animals. Many of the acts were similar. Turn-of-the-century animal performances at zoos, however, were generally genteel compared to those at circuses, as well as those at amusement parks and the midways of fairs and expositions. At the Bronx Zoo and others, visitors could observe the good manners of the orangutans as they took their daily tea, for example. Other animals outside their cages were similarly tame—zoos operated profitable sidelines in rides on the backs of elephants, camels, and tortoises. Detroit's brandy-swilling, cigar-smoking chimpanzee Jo Mendi, a veteran of the RKO vaudeville circuit and Broadway, was perhaps the most vulgar of zoo performers. Still, he danced, walked the tightrope, rode a bicycle, and roller skated—acts hardly likely to offend zoo audiences in the 1930s. The St. Louis Zoo was also well known for its performing chimpanzees and other animals.[42]

But despite the overlap between zoo performances and other animal shows, there were important differences. Zoo shows pale in comparison to other places where exotic animals were displayed, in terms of both their scale—the numbers of animals and trainers—and the kinds of stunts involved. A late-nineteenth-century competition among circuses for elephants brought twenty-five elephants into the ring of the Adam Forepaugh show, whereas a large zoo considered itself lucky to have two or three elephants. At Coney Island's Luna Park the pachyderms slid down a special elephantine Shoot-the-Chutes. Luna also featured the diving horse King, who leapt from a platform into a tub of water.[43]

The Coney Island animal shows derived from attractions at the midways of world's fairs. These too were flamboyant beyond the means or goals of most zoos. At the Buffalo and St. Louis world's fairs "educated" horses did arithmetic, picked up playing cards, and counted coins. At the 1901 Buffalo Pan-American Expo and others, Frank C. Bostock's Arena of Trained Wild Animals featured more than a thousand "caged beasts" as well as an act with twenty-five lions. In the amusement area of the 1939–40 New York World's Fair, Frank Buck's Jungleland advertised 30,000 animals and birds, including 1,000 rhesus monkeys displayed on a concrete monkey mountain. Even allowing for exaggeration, this was far more than the 2,000 to 3,000 specimens kept at the largest zoos.[44]

World's fair midways also portrayed human-animal relationships in ways that zoos did not. At fairs, animals often formed part of ethnographic displays—Laplanders with their reindeer, Eskimos with their sled dogs. Camels wandered the "streets of Cairo," and "Hindu" snake charmers performed with cobras. At the 1933 Chicago fair, "the only attraction of its kind" was the alligator farm, where, according to a guide book, "[T]wo Seminole Indians plunge into a pool and, after an exciting struggle, manage to land a 375-pound saurian on a platform."[45]

By contrast, American zoos and their audiences generally considered it inappropriate to display humans. There were exceptions: the Bronx Zoo briefly exhibited an African Pygmy named Ota Benga in 1906. The zoo ostensibly employed Benga as a keeper for a chimpanzee. Vocal protest from New York's black community, however, cut Benga's zoo career short. Native Americans were occasionally enlisted to add interest to zoo displays as well. But rather than showing animals as part of a network of human-animal social interaction, zoo displays more commonly recalled human culture through architecture, by putting animals in houses that resembled the building style of the creatures' country of origin. In keeping with this architectural associationism, zoos exhibited American bison in log cabins. Ethnographic displays were not the mainstay of zoo exhibits, unlike those at fair midways.[46]

Above all, what distinguished animal shows at zoos from those at other places was that zoo shows took place in the landscape of a park. The zoo stage was in a landscaped setting of winding paths, and it juxtaposed middle-class leisure pursuits with the experience of looking at animals. At many zoos visitors could rent rowboats or listen to band concerts. Golf courses and tennis courts were often located in the parks where zoos were situated, although separated from animal displays by fences and hedges. At the zoo, as opposed to the rowdy amusement park, there was no Steeplechase, no "Cannoncoaster," and no "Human Whirlpool." Zoos had no hootchie-kootchie dancers, midgets, reenactments of the Johnstown flood, or "Coontown plunge" concessions. Historians have discussed amusement parks as landscapes of rebellion, where working-class crowds challenged genteel culture. Zoos did not fit this description. They were middle landscapes—landscapes of peace, harmony, bounty, and wonder—infused with a middle-class ethos and intended to elicit middle-class standards of behavior.[47]

When turn-of-the-century zoos are compared to world's fairs, they resemble formal exposition areas rather than the midway in their landscape and ethic. Most zoo directors involved with day-to-day animal keeping shunned "fantastic designs" and "gaudy colors," considering them "undignified and inappropriate." At the National Zoo, "[s]imple, substantial buildings, with little ornamentation, but with pleasing outlines," seemed most suitable. A 1928 manual on park planning similarly recommended "plain and dignified construction" for zoo buildings. Despite such rhetoric about unobtrusive animal houses in rustic zoo parks, however, when they had the funding to do it, zoo planners built animal houses that represented the cutting edge of both style and science in animal keeping. The role of zoos as civic institutions sometimes compelled them to erect monumental animal houses. Like fairs, zoos could fulfill their mandate to educate—to elevate public appreciation of wildlife—and to set standards of culture, through architecture.[48]

To cite a specific example, the World's Columbian Exposition of 1893, with its much-emulated Beaux Arts "White City," provided an architectural model for the Bronx Zoo. Designed by architects George Lewis Heins and Grant LaFarge, the 1897 plan for the zoo's central cluster of buildings is similar to the Columbian Expo's Court of Honor. Animal and administration buildings surround a long narrow rectangle of flat open space, occupied by gardens and a sea lion pool. One end of the courtyard is open; the domed elephant house, reminiscent of the 1893 administration building, caps the other end. At the zoo's opening ceremonies, the prominent Zoological Society member Henry Fairfield Osborn explicitly referred to the congresses on literature, history, science, and art at the Columbian fair; he declared the collection of animals "A World's Congress of Lions."[49]

The Landscape of Progress

The White City of the Chicago fair, as is well known, became a model for city planning at a time when cities were expanding and there was much rivalry among them. Zoos became esteemed and desirable additions to plans for city expansion. Zoo directors had admired the high social status of zoological parks in Europe and the prominent role zoos played in civic culture, and they hoped that American zoos would fulfill a similar role. Hornaday observed that in Europe, "It is considered an honor as well as an advantage to belong to those zoological societies." Zoo promotors compared their animal collections to institutions such as museums. According to the New York Zoological Society, "The movement to establish a great zoological park for the people of New York, is the outcome of the civic spirit which has established the Museums, the Public Library, and the Botanical Garden." With characteristic hyperbole, Hornaday declared the zoological garden "the high-water mark of civilization and progress." In smaller cities as well, a zoo was seen as an obvious accompaniment to an expanding population. "It is the desire of our city to organize a zoo," wrote a resident of Wichita Falls, Texas, in 1926. He explained: "In the last eight years, we have grown from a town of 8,000 to a city of 60,000. Much interest has been shown in contributing to the upbuilding and improvement along any line."[50]

In other cities, advocates touted zoos as the pinnacle of park design and a tourist attraction. One Harvard paleontologist declared that "the city park, if developed to its highest power, should give the necessary space for zoological gardens." In any case, a zoo was a badge of rank for a city, a sign of its significance. Members of the Million Population Club of St. Louis (population 700,000) hoped that a zoo would attract new residents to their city. Evansville, Indiana, built a zoo because the city "was not getting the recognition, the trade, the good will or the travel" that the Metropolis of the Tri-State "should reasonably expect." Duluth, Minnesota, considered its zoo "a fine reflection on the spirit and stability of the city." And the mayor of Memphis—a city without a zoo in 1901—confessed to the director of the National Zoo: "Memphis has not yet progressed sufficiently to have a collection of animals in its Park. We hope, however, for this at no distant day."[51]

With such civic support, which usually included some form of municipal funding as well as a site in a public park, zoological parks had indeed set themselves far apart from circus menageries and animal vaudeville performers. Zoo planners accomplished this cultural transformation in large part by adopting a geography of gentility for their displays, a landscape recognizably middle class in its location in a park, and its resemblance to suburbs and campuses. This landscape of learning, and the elevated social status accompanying it, helped zoos to pursue their goals as educational institutions. To be sure, tension remained over whether zoos were more than mere entertainment— some members of Congress, for example, initially ridiculed the idea of a National Zoo, saying that it would rival Barnum. But whether zoos were educational was not an issue in the eyes of the public schools, which sent children on zoo trips early on. The Cincinnati public schools began sending tens of thousands of students on regular zoo visits in 1896. In the 1890s, the Philadelphia Zoo donated ten thousand free admissions to the board of education to be distributed to students, beginning a long-term relationship with the city's education system. The Bronx Zoo boasted that more than twenty-three thousand children had visited the zoo with their teachers in 1909 and 1910. The San Diego Zoo operated a bus to transport children whose visit was coordinated with the nature study department of the public schools. The San Diego Zoo was also among the first to offer college-level training, with its "zoo science" extension course coordinated through San Diego State College in the 1930s.[52]

Although it is not always clear what education meant to zoo proponents—and clearly it was different things to different people—all agreed, along with park planners, that exposure to nature was in itself beneficial. Everyone could profit from a visit to the zoo: young, old, rich, poor, scientist, teacher, student, "even the mere gazers." The animals would both serve as models for artists and taxidermists and entertain toddlers. Most of all, however, zoos promised the combination of education and recreation, a place where school children could go and "absorb much valuable knowledge of wild life without effort on their part, a knowledge which they retain throughout their lives."[53]

The metaphor of Noah's Ark has long been irresistible to zoo planners; a breeding pair of every species would make not just a complete collection, but one that would perpetuate itself, sparing the zoo the future expense of buying animals. Clearly, no zoological park could accommodate specimens of each of the hundreds of thousands of known species, and in fact zoo directors never attempted to bring their collections to a biblical state of completion. In the late nineteenth and early twentieth centuries zoo directors often said that they aimed to exhibit a representative sample of the world's fauna. They focused on vertebrates—mammals and birds especially. A representative collection, unlike the ark, was finite and obtainable. Showing complete series of related species was not an important goal. For the numerous rodents, William T. Hornaday readily admitted that "it is impracticable to do more than place before visitors a reasonable number of well-chosen types, which shall represent as many as possible of the twelve families, and also the genera most worth knowing." A representative collection would contain enough animals from each order, displayed near each other, for visitors to observe how the forms of the various carnivores, for example, were similar and where they diverged. It also needed enough variety to be authoritative—for visitors to understand the zoological park as standing for the world of wild animals, for them to believe, as one zoo director put it, "We give you nature."[1]

Hornaday's allusion to animals "most worth knowing" is a tip-off to the fact that the Linnaean system of classification was not the only guide to selecting animals for a zoological collection. Deciding who belonged in a zoo collection was more than a process of abstracting a list of the world's fauna. Zoo directors set priorities; one of Hornaday's goals was to display bison and other American big game. And city park administrators who had no special knowledge of animals could read articles in their professional journal, *Parks & Recreation*, that laid out the range of species they should seek for their zoo. Practical considerations also played a role in selecting animals for zoos. As adjuncts of city parks departments, most zoos had small budgets for operations and even less for purchasing stock, so they exhibited deer and birds that were inexpensive and easy to keep.

Furthermore, zoo visitors arrived at the gates with their own ideas about what they expected to see on display in zoos. Popular notions about who belonged in the zoo were based on previous exposure to animals in many contexts: performing animals in circuses and vaudeville, pets and farm animals, hobbyists' fancy breeds, the bats, squirrels, and snakes that they found in their backyards. This well-informed audience contributed quite literally to assembling and ordering zoological collections. Wherever new zoos were built, an enthusiastic public deluged them with donated animals and offered others for sale. When the National Zoo opened in 1889, American naturalists in the West sent bears and prairie dogs. Military officers, ship captains, and government zoologists stationed overseas contributed warthogs, antelope, lions, and tortoises. Other members of the public offered still more unusual specimens: a three-legged chicken, a calf of ambiguous sex, and an albino squirrel. In addition, the National Zoo—like other zoos—received donations of rabbits, rats, mice, and various unwanted pets. Animal donations were most often made by individuals, but in several cities citizens rallied together to collect money to buy an elephant—an animal too expensive and unwieldy for most individuals to manage.[2]

Zoo directors had mixed feelings about these popular contributions. Donated animals usually were not high on the list of species they considered desirable for display. But small budgets forced them to keep an open mind; after all, it was always possible that some local zoo lover traveling overseas would pick up a valuable specimen. Even so, the steady stream of birds, bats, opossums, and rabbits deposited on zoo doorsteps tried their patience. Frustrated by "not being able to refuse the numerous small things that people would thrust upon the zoo," the Secretary of the Smithsonian Institution, Samuel Pierpont Langley, asked the director of the Philadelphia Zoo, Arthur E. Brown, for advice. Brown told Langley that "he must purchase a big snake, which would devour those things rapidly. He (Mr. B) had had an alligator that eat [*sic*] up all the small ones that people flooded the gardens with. An animal with a big appetite was a very desirable addition to the zoo."[3]

Zoos were new places, civic institutions that appealed to diverse audiences whose ideas about which animals belonged often differed from the representative collection to which zoo directors aspired. People who donated animals participated in their local zoos outside of organized forums such as ladies' auxiliaries and junior zoological societies for children, which parroted the goals of zoo organizers. In their letters to zoo directors, they described the creatures they wished to donate or sometimes sell, how they came to possess the animals, and why the animals belonged in the zoo. Indirectly, they expressed their ideas about what sort of institution a zoo should be, which animals were appropriate for display in a zoo, how the human–animal encounter within the zoo should be constructed, and how they ordered the natural world.[4]

Even though zoo directors often refused offers of animals from the public, the record of popular contributions to zoos offers a way to understand the public negotiation about what sort of place a zoo should be. Whether their local zoo was managed privately or by city government, many people perceived it as their civic duty to make a contribution. They felt they had knowledge and expertise to add to the zoo. While theirs was not exactly a system of classification to rival that of Linnaeas, the zoo-going public superimposed on the taxonomic order additional categories important to assessing a zoological collection's completeness, identifying animals at the extremes of exotic and mundane, ferocity, tameness, and size. Some people who donated animals understood the function of the zoo in similar terms as zoo directors. Their idea to collect animals came from a tradition of popular natural history collecting, and a wish to participate in a scientific institution. Other people conflated zoos with places like circus sideshows that also exhibited exotic animals. Another segment of the public perceived the zoo as a sort of animal hotel, in which they could deposit exceptionally well-behaved, intelligent, or accomplished pets. To these people, the animals in the zoo were members of an extended human-animal family. People who organized campaigns to raise money to purchase large animals for their local zoo held yet another concept of the zoo's purpose. In Boston, for example, a campaign to purchase elephants for the zoo became an effort to develop community ties across class boundaries through a project for the city's children. The zoo became a symbol of civic unity.[5]

Despite this range of interpretation, two things seem to have been generally agreed on in the first decades of American zoos by zoo directors and their publics. The first was that common farm animals did not belong in a zoo, except perhaps a pony for carrying children around a track. There was the occasional exhibit of useful animals or insects such as honey bees. But poultry, hogs, cattle, and so on, which would have been displayed at state and county fairs, were never considered appropriate for zoos. This was so obvious to people at the time that it was never discussed. Displays of farm animals certainly could have been educational; perhaps they were merely too familiar, not an attraction to an urban audience that relied on horses for transportation and could raise their own hens within the city limits. Only in the 1940s did Americans begin building children's zoos, petting zoos, and farms-in-the-zoo, as populations of both urban domesticated animals and rural humans decreased.

The second widely understood notion about zoo collections was that a zoo was not a zoo unless it had an elephant. The elephant was the keystone of the collection. Most zoos considered the day that their first elephant arrived as the day they became "real" zoos. Milwaukee's zoo, for example, had been founded in the early 1890s. But only in 1906, when the Knights of Pythias donated an elephant, was the zoo changed "from a joke into a serious matter for general public consideration." The elephant was "the injection of elixir required to breathe the breath of life into a movement that had languished, hovering between life and death, for years." Recounting the origins of the National Zoo, William M. Mann wrote that two donated elephants were "marched out from the circus grounds" to the zoo, ". . . chained to a tree, and the zoo was a fact."[6]

In between the obviously unsuitable barnyard chicken and the essential elephant, among the bears, tigers, pet monkeys, prize-winning goats, rare birds, and dogs, there was room for negotiation about who belonged in the zoo. Public participation in the zoo through animal donations reveals a wide range of relationships and historical associations between humans and animals, and popular donations helped make the zoo a civic institution with which people felt personal ties. In turn, responding to their public forced zoo directors to articulate their interpretation of the zoo collection.

The Desires of Zoo Directors

From the start, the idea of a zoo as a sort of living museum of type specimens from around the world was tempered by practical considerations. Zoos had to make do with the animals available. Some species of exotic animals were available for purchase from commercial animal dealers; but assembling a collection through purchase was expensive, and exotic animals did not arrive in American ports with the same regularity as in London or Amsterdam in the nineteenth century. Comparing their collections to those of European zoos, American zoo directors recognized that "here, we are handicapped by lack of colonial possessions." As an alternative to tropical birds and unusual antelope, they discussed the merits of focusing on North American animals.[7]

For purposes of education, some zoo supporters believed that displays of local fauna would be both practical and suitable. An early supporter of a zoo for Boston suggested that "when we remember that not one in ten thousand, perhaps not one in fifty thousand, of our city people (not only here in Boston, but anywhere), has ever seen or is in any way familiar with the greater part of the animals and plants that are indigenous to the soil on which he was born and bred . . . we see at once that we have here an opportunity of setting an example to the world, sure to be followed to the gain of general education everywhere."[8] A collection of "New England indigenes" would have further advantages. Such animals would be easy to obtain and would require minimal housing, and no one would confuse the zoo with the circus menagerie: "elephants and giraffes, camels and tigers would not be expected."[9]

With more ambitious goals than the Boston zoo supporters, the National Zoological Park and the New York Zoological Society's Bronx Zoo both began with the intention of displaying and breeding big game animals native to North America. In its first formulation, the National Zoo was to be a "city of refuge" for bison and other North American mammals that were in danger of extinction by hunters. Likewise, the wealthy sportsmen who founded the New York Zoological Society hoped to create a park where such species could roam and multiply, eventually building a population that would be reintroduced to the West. The commitment to conservation was unusual for American zoos at the turn of the twentieth century, however, and it formed only a part of the mission of the zoos in Washington, D.C., and New York. Although large North American mammals—including bison, elk, moose, and mountain sheep—certainly belonged in zoo displays, they were not enough to make a zoo. Animals that evoked national pride were just one class among others at the zoo.[10]

Large mammals were not the only important American species. At the Bronx Zoo, director William Hornaday proposed displays of native wolves and foxes. Even those smaller animals often taken for granted were appropriate for the zoo. He wrote: "The collection and arrangement of American rodents, both burrowing and arboreal, will,—for perhaps the first time,—do justice to the splendid series of forms of this order which are native to our country. It is probable that very few persons, outside the ranks of our own mammalogists, are aware that our country possesses the greatest variety of squirrels and marmots to be found in any one country, and that the most beautiful forms are the ones most seldom seen."[11] Although they may not sound spectacular, such exhibits were well received. Every zoo counted a prairie dog "village" among its most popular displays. Prairie dogs, although small, were beloved for their sociability. They were easy to anthropomorphize and reminded visitors of the relatively recent expansion of European-descended Americans into the western United States.[12] Hornaday described both the prairie dog's appeal and its status as a pest: "Owing to his optimistic, and even joyous disposition, the Prairie-Dog has many friends, and "happy as a Prairie-Dog" would be a far better comparison than "happy as a King. . . . His so-called bark is really a laugh, and his absurd little tail was given him solely as a means of visible expression of good nature. But he has his enemies and detractors. The coyote loves his plump and toothsome body; the 'granger' hates him for the multitude of his holes, and puts spoonfuls of poisoned wheat into his burrow."[13]

Animals native to a zoo's locale, or to North America, were relatively easy to obtain and to keep. Experienced directors advised new zoos to begin by displaying these animals, proceeding with "gradual development, as interest increases and funds are available. You might even start without exhibition buildings, or heated quarters of any kind, by construction of paddocks and shelters for hardy deer, bison, and similar animals. Some species of monkeys, lions, wolves, etc., could also be exhibited . . . without artificial heat."[14]

The well-planned, gradual approach to zoo expansion was usually scrapped, however, as zoo directors took advantage of any animals that were readily available, especially if they would attract a crowd. The Denver park board, for example, gave a commercial game breeder permission to exhibit his exotic pheasants in the zoo. As a result, the zoo could boast "the largest collection of Chinese pheasants in the country" and an inventory of three thousand specimens. Circus animals often spent the winter in city zoos, at least temporarily adding out-of-the-ordinary specimens to zoo collections. The National Zoo even considered an exhibit of domestic dogs, for the educational purpose of showing variation within a single species. Not least, however, enthusiastic zoo visitors also did their part to increase the animal population of the zoo. Their donations both boosted collections and forced directors to place boundaries on which animals were appropriate for display.[15]

Amateur Naturalists

The rush to donate or sell animals to new zoos reflected in part the wider popularity of collecting and trading natural history objects. By the late nineteenth century, thousands of Americans—men, women, and children—collected natural history specimens in their leisure time.

They arranged their shells, bird eggs, fossils, flowers, minerals, and mounted animal skins in cabinets at home. They bought, sold, and traded specimens among friends and through advertisements in magazines such as *The Museum*. The 1879 edition of the *Naturalists' Directory*, a commercial listing of individuals engaged in natural science, published more than three thousand entries. In addition, books for boys provided advice on how to collect bird eggs and nests, as well as lessons on practical taxidermy. Amateur naturalists who collected animals for zoos shared this passion for collecting.[16]

The natural history community of the nineteenth century comprised a broad range of education and ability, and categories of amateur and professional fail to characterize it adequately. Amateur natural historians could make important contributions to scientific knowledge, and so-called amateurs and professionals depended on each other's work. Ornithology, for example, remained incompletely professionalized as late as the 1930s. Museum curators who studied birds relied on people who did not work in museums, universities, or the government, and who were not primarily ornithologists, to collect birds and to record observations on bird habits and migrations. At the Smithsonian Institution, Spencer F. Baird cultivated a widely distributed network of collectors in the nineteenth century; half of his correspondents came from towns with populations of less than four thousand, and about twenty percent came from southern and western states. Baird's correspondents sent specimens to the Smithsonian's National Museum. In return, Baird sent them books, as well as apparatus for collecting, preserving, and labeling specimens. Contact with the Smithsonian also helped isolated collectors maintain their enthusiasm, and it drew public attention to their work. The Smithsonian's correspondence network declined in the 1880s, after Baird assumed heavier responsibilities as the institution's secretary, but amateur naturalists continued to perceive the Smithsonian as a place that would welcome their contributions of specimens.[17]

The National Zoo, as a branch of the Smithsonian Institution, benefited from the Smithsonian's reputation. Amateur and aspiring naturalists from the Washington, D.C., area and across the country regularly deposited their prizes there. Opossums seemed to be a favorite catch of the backyard naturalist, as did squirrels, skunks, raccoons, hawks, and rattlesnakes. William H. Babcock wrote to the National Zoo, "My boys have developed a mania this summer for catching things." He wanted to know if the zoo would encourage their interest in the natural world by buying their turtles and snakes. After a visit to the zoo, F. H. McHaffie of Wevaco, West Virginia, decided that it "should have a good Rattle Snake from the head of Cabin Creek"; he proceeded to catch three and send them along. Other contributors came upon wild animals by accident rather than intent, and shipped them to the zoo. Mrs. J. H. Cummings of Wilmington, North Carolina, sent to the National Zoo "a bat and her four babies. She was found by Mr. Stevens, a guard over the road gang, working on road in front of our place." Similarly, the National Zoo received a letter regarding owls stating, "Mr. Joseph W. Hawkins, Janitor of 'the Alabama' caught these owls in the Tower of the Temple Baptist Church and desires them presented to the National Zoological Garden."[18]

The National Zoo depended on such donations. Always subject to the whims of congressional appropriations, early in its history the zoo was granted no budget at all to purchase animals. According to the zoo's annual reports, it received between eighty and one hundred donated animals per year in the decades around the turn of the century. (The zoo also acquired animals through trades with other zoos and through "births and hatchings.") Although grateful for the animals that he could get, the zoo director acknowledged that this system had its drawbacks. "The animals given are, it is true, sometimes very valuable," he wrote. He added, however, "They are usually the random, accidental finds made by chance sportsmen or curiosity hunters, and are, naturally, more numerous in some classes than others. Numbers of opossums, raccoons, and small alligators are yearly presented, but no one has ever thought of presenting a moose, a caribou, a manatee, a sea-lion."[19]

As a part of the federal government the National Zoo also could tap into a source of collectors outside the United States. Among diplomatic and military personnel stationed abroad, the National Zoo found amateur naturalists eager to collect animals. In an effort to cut down on random gifts and to direct the efforts of such government employees toward its goals, the Smithsonian Institution distributed a pamphlet titled *Animals Desired for the National Zoological Park* in 1899, similar to

manuals issued by museums for collectors of natural history specimens. The circular listed species desired according to continent of origin, and gave instructions on how to feed, crate, and ship them. From
South America, for example, the zoo especially hoped to receive a
tapir, sloth, anteater, great armadillo, or jaguar. Cooperating with the
United States Secretaries of State, of War, and of the Navy, the Smithsonian distributed the guide to military officers and diplomats stationed in the Philippines, Cuba, Puerto Rico, Hawaii, and elsewhere.[20]

The call to collectors elicited many prompt responses: the governor
of Puerto Rico, for example, sent a yellow-billed curassow. The U.S.
Consul in Maracaibo, Venezuela, was enthusiastic, and was typical of
these overseas collectors who were observant of the local fauna and
perhaps bored with their desk jobs. He offered to send "tapirs, young
tigers, deer, eagles, 15 feet long, live aligators and anoconda [sic] of
over twenty feet long, birds and monkeys of all description." He re-
searched a way to secure free steamship transportation for the animals
to New York. "I hope that the next congress will allow a liberal amount
of money to the National Zoological Park to enable Consular Officers
to go to work systematically and buy up and collect whatever is
wanted," he added, offering his own services. "I know that should a
few hundred dollars be allowed to me to buy up and collect specimens
I could give full satisfaction and be of great benefit to the Institution.
Anyhow I will do my best as far as my own slender means allow me
to do."[21]

Although the National Zoo's contacts were exceptionally well developed, many zoos had access to such collecting networks on a smaller
scale. A zoo director might draw on friends in university zoology departments, such as Harvard zoologist Thomas Barbour, who regularly
donated animals to the Philadelphia Zoo and the National Zoo. A zoo
director might also have contacts from an earlier career as a field scientist, like Edmund Heller, the naturalist who accompanied Theodore
Roosevelt to Africa in 1909 and later ran the zoos in Milwaukee and
San Francisco. John Alden Loring, the Denver Zoo's first director, had
been a naturalist with the U.S. Biological Survey. The New York Zoological Society's membership, which overlapped with the Boone and
Crockett Club, a group of wealthy hunter-naturalists, contributed to
the Bronx Zoo's animal population. Other zoos had access to collectors

ADVICE TO COLLECTORS.

ANIMALS ESPECIALLY DESIRED.

The new possessions of the United States are comparatively poor in animals, but it is especially desirable to have as full a representation of the fauna as possible. While all will be valued, those whose names are italicised are particularly desirable.

CUBA AND PORTO RICO afford the *manatee*, or sea-cow, which frequents bays and mouths of rivers; the *flamingo*, spoonbill, ibis, pelican, several species of parrots and parrakeets, a variety of pigeons, the ani, and other interesting birds. Boas of several kinds occur in these islands, and large lizards of different species are very abundant. The agouta (*Solenodon*) and the hutia (*Capromys*), animals a little larger than a common rat, and the crocodile are also found in Cuba, and an interesting macaw occurs in the Isle of Pines.

In the PHILIPPINE ISLANDS the most notable mammals are the

"tamarau," a small wild buffalo found on Mindoro, several species of deer, the "babui," or wild hog, monkeys of two species, a small cat, two species of civet cat, or musang, fruit-eating bats of different species, several peculiar large rats, the *colugo*, or *flying lemur*, and the very remarkable and interesting *tarsier*, or "*magau*." Among these the last two and the "tamarau" are especially important. Specimens of the domesticated buffalo also are desired.

Of the birds, the eagles, hornbills, cockatoos, parrakeets, the pheasants and pigeons, the megapod, pelican, and the ground cuckoos are perhaps the most important. Among these, any *hornbills* or brilliant-plumaged *cockatoos* and *parrakeets*

FIGURE 7. The National Zoo issued a brochure giving guidance to potential collectors. From: *Animals Desired for the National Zoological Park at Washington, D.C.* (Washington, D.C.: Government Printing Office, 1899).

through the natural history societies and museums with which they were affiliated.[22]

Even without previously established collecting networks, and without soliciting, new zoos received donations. The St. Louis Zoo at its opening, according to a newspaper article, "attracted the attention of the entire nation, from every section of which offers to sell or give animals, birds, or reptiles are being received." Indeed, the zoo's annual reports acknowledge dozens of donations, mainly from people with St. Louis addresses.[23]

Like the far-flung collectors in Spencer Baird's Smithsonian network, people who donated animals to the National Zoo could feel they had a hand in building an important cultural institution. They got credit for their work; the National Zoo, like other zoos and also museums, printed the names of animal donors in their annual reports, alongside the species given. The National Zoo's director answered letters from the public personally. He offered tips on caring for the animals offered, and often sent shipping tags and crates and paid the cost of transporting the animals. More than other animal donors, amateur naturalists' conception of who belonged in the zoo was similar to the ideas of zoo directors. They saw it as a place to display normally developed specimens caught in the wild, some of them common, others unusual. They offered donations in order to further the zoo's function as a scientific and educational institution. From their perspective, the zoo was a place to learn by observing living animals, and they were proud to make their contributions to this public good.

Freaks of Nature

The National Zoo also received dozens of letters from people who tried to persuade the zoo directors to take—preferably to buy—albino, hybrid, sexually ambiguous, or deformed animals. "I have the greatest living curiosity of the age," wrote G. W. Armistead to Frank Baker, the director of the National Zoo, in 1897: "It is a calf 1 1/2 years old, small for the age. To raise its tail and examine thoroughly one would vow it to be a male, but when you examine for the penis one would vow it to be a female. . . . I expect to make big money for this animal." Letters offering freaks of nature like those commonly exhibited in dime museums and circus sideshows identify a segment of the public to whom zoos represented just one among many arenas displaying live animals at the turn of the century. They saw zoos as akin to places that displayed unusual animals to a paying public. And they could hardly be blamed for this interpretation. For decades, showmen of various persuasions had exhibited an odd assortment of live animals without distinguishing between those rarely seen, such as giraffes, and others with genetic or developmental deformities.[24]

Furthermore, P. T. Barnum and others in the nineteenth century had borrowed the language of scientific description and offered expert testimonials in their popularizations of natural history. Sometimes credible, sometimes humbug, it was an effort to convince the public of the educational value of their animal shows, and to attract a respectable clientele. In the first few decades that zoos existed in the United States, some people perceived them as just another variation on this familiar theme, and they expected to find the same curiosities within. The National Zoo's correspondents seem to have interpreted the zoo's professions of scientific seriousness as an effort to attract a well-heeled audience. In response, they mimicked both the language of natural historical description and the entrepreneurial spirit of dime museums, circuses, and vaudeville shows. H. C. Baird, of Tibbee, Mississippi, for example, described the anatomy of his three-legged chicken with a naturalist's attention to detail: "It is a yellow pullet over half grown and hardy. . . . The third leg is behind and hangs down like the tail of a dog or hog. The leg is perfect in shape but is connected to the body by flesh and not by bone, but the third leg has the full bone in it. also toes perfect, and that leg is feathered like the other two leges [sic]." M. D. Oakly, of Oxford, North Carolina, offered a rooster with four wings, "three on one side and one on the other." He had no doubts about what to do with such an animal: "Knowing this to be a curiosity and being anxious to sell it, I write to you."[25]

Such entrepreneurs also seem to have confused the zoo's interest in breeds (the zoo maintained an exhibit of dog breeds early in the century) and species with their own interest in unusual genealogies and abnormal development. They wrote to the zoo director with albino animals: hawks, crows, woodchucks, raccoons, and hedgehogs. They offered wolf-dog hybrids for sale. Additional animals perceived as appropriate for the zoo included a hairless mare, "a little goat . . . with nice white wool, two perfect heads and necks, five legs, and two tails," a sheep with "a perfect horne on the end of Nose just as a Rhinoceras [sic]," a six-legged dog, an earless hog ("if you know any show people you think would like to buy it please give me their name and address"), and "the greatest wonder of the age. . . . A horned rabbit."[26]

The zoo director at the time, Frank Baker, responded patiently, expressing gentle skepticism and explaining to letter writers the difference between the freaks they offered and what the zoo wanted. Answering the description of "a family of cats whose great grandfather was a jack rabbit the mother a maltese cat," Baker wrote, "Animals with the ancestry which you mention would certainly be most unusual, as the interbreeding of animals so far removed from each other zoologically is rare indeed." He reminded others that his purpose for the zoo was "to include in the collection only specimens which show the normal development of the species to which they belong" and "to exclude from our collection animals which are merely freaks or curiosities."[27]

People who offered freaks to the zoo focused on the zoo as a place for entertainment, similar to other venues that displayed exotic live animals. They perceived animals as commodities to be sold to zoos. Their letters also indicate that they interpreted the zoo as a profit-making institution that would be interested in animals that drew a paying crowd.

A Garden for Pets

People who thought of the zoo as a sort of boarding house or hotel for an elite class of animals provided another source of donations to the zoo. To these people, sending a pet that could not be kept to a zoo was like trusting the care of a child to relatives. In trying to convince the zoo to take their pets, they boasted of the animals' exemplary behavior. At the zoo, pets would become acquainted with their animal cousins, and would be well fed. Sending a pet to the zoo was a difficult decision for many of these people, a decision they rationalized by imagining the zoo as a place that would be healthy for their pets and where their pets could bring pleasure to others.

Some members of the zoo public wished to board their pets, as in a kennel, while they went on vacation. Elizabeth A. Arnold, for example, wrote to the National Zoo concerning, "a parrot which I desire to have kept at the Park for the rest of the summer." The cadets of the U.S. Coast Guard Academy in New London, Connecticut, left their bear at the National Zoo during their cruise. A pet's stay at the zoo was often looked at as a sort of privileged summer camp, temporary accommodations that required references for admission. In attempting to persuade the zoo to care for a monkey, its owner described it as "healthy, gentle, and well disposed." Zoo directors did caution such people that their pets might not be as refreshed as they by the break from routine. William M. Mann warned the monkey owner that "we should be very glad to have the monkey here. However, when a pet animal is put into a zoo it usually ceases to be a pet. It does not get played with as it does in a house, and sometimes it is difficult to make it tame again when you get it home."[28]

Nonetheless, just as it took in circus animals as temporary boarders, the National Zoo also accepted some pets, adjusting its policy on this practice depending on the species of animal offered and the space available in the animal houses. The zoo turned down a child's request in 1903 to board five pigeons on the grounds that the bird house was full. But in 1915 it took in a parrot ("which is a whistler and talks quite a little") from an engineer who went to Australia frequently on business and had donated Australian animals to the zoo. It was worth currying favor with such a well-traveled and generous person. Head keeper William Blackburne observed, "he might be of some help to the Park in the future during his travels. . . . It might be best to care for the Parrot."[29]

Rather than a temporary home, however, a larger segment of the public turned to the zoo as a dumping ground for pets they could no longer keep: souvenir alligators from Florida that grew too large, wild cats that had been cute as kittens but presented a danger to children when older, animals that children got bored with and parents did not care to bother with, and animals that people could no longer afford to keep. Various clubs found themselves overwhelmed by their hungry mascots: the Moolah Temple of Shriners gave the St. Louis Zoo a camel; in Philadelphia, the Loyal Order of Moose, Philadelphia Lodge

No. 54, donated a "fine bull moose." Military battalions deposited their mascots in zoos as well.[30]

Such animal donors did not see the zoo as just a stockyard, however, where pets would receive minimal attention. They considered the alternatives—selling the pet, giving it to someone else, or having it put down—and decided that the zoo represented a humane solution, a grassy place where their pets would be taken care of and would serve the public. Mrs. Long, in Fallon, Nevada, was "very glad" that her pet eagles went to the National Zoo, where they would have "a permanent home and where they will be well cared for." Mrs. C. M. Buck of Washington, D.C., was too attached to her double yellow-headed parrot to sell it; rather she wanted to "contribute him to The National Zoo to be retained there as Government property and exhibited for the benefit and pleasure of all."[31]

Like those who sought temporary room and board for their pets, people who deposited their animals permanently at the zoo described the qualities that they believed would make their pet an asset to the zoo. Often they emphasized how their animal was healthy and docile—but active and interesting to watch, well behaved, and appropriate for children. A parrot had "a wonderful vocabulary, saying many things pleasing to hear and nothing one would not want children to hear." A coyote was "as tame as a dog, and has never attempted to bite." A fox aptly named Reddy had "more than his beauty to recommend him," being "clever, active, docile, and playful as a kitten." An American wild cat was "as thoroughly gentle and tame as a domestic cat," but unfortunately had "taken to killing the neighbor's chickens," wrote J. A. August of Pine Hill, Kentucky, adding, "The animal's name is Beauregard."[32]

Other pets deserved sanctuary in the zoo because of their exemplary achievements. Given that many zoos displayed animal celebrities that had been donated, this must have seemed reasonable. The Philadelphia Zoo exhibited the "Eskimo sled dogs" that had accompanied Robert Peary on an expedition to Greenland, for example, and an all-black wolf in the Denver Zoo was nationally known for his heroic deeds. In a similar vein, a letter writer to the National Zoo offered a Colorado burro known as Jerry, who had "quite a good reputation and an interesting career, having played on all the stages of the leading theatres in

FIGURE 8. The grave of the Denver Zoo's beloved "Black Jack, Wolf Wonder." Denver Public Library, Western History Department.

Washington, and at various times loaned to prominent actors." C. A. Beard, of Rockville, Maryland, sought "somebody who will take good care of" his prize-winning goat. Genevieve B. Wimsatt hesitated to leave her well-educated Chinese bird "with bird dealers, who might neglect it," so she elaborated on its abilities for the National Zoo: "The bird readily imitates the notes of other birds, the squeaking of wheelbarrows, and other sounds, and can be taught a few simple words. . . . It articulates quite plainly a few phrases in the Southern Chinese dialect, as well as a bar or two of a Chinese song. During its stay in Tientsin it learned to call a rickshaw 'Chiao P'i! Chiao P'i!' also to say, 'Keep quiet! Keep quiet!' It also says, 'What?' and 'Edgar!'"[33]

Occasionally pet owners had second thoughts about abandoning their charges. The monkey Mike Stiles was deemed too destructive to the florist shop of its owner in Petersburg, Virginia, and sent to the National Zoo. A year later she wrote to ask if she could have Mike back, explaining, "I have a little girl here who is grieving her self sick about that monkey." More often, pet owners merely wanted to visit their pet in the zoo. Fannie M. Hawkins hoped to visit her pet opossum Snooks, and H. N. Slater, who gave up his white baboon to the National Zoo, wrote, "The next time I am in Washington I shall be anxious to go out to the park to see her."[34]

Articles in newspapers and magazines perpetuated the idea that zoos provided luxury hotel service to their inhabitants. Newspapers served as welcome wagons for new additions to local zoos, introducing them as neighbors and providing information on where the animals came from and what they liked to eat. According to magazine articles with titles like "How Jungle Beasts Live Near Our Big Cities," "Backstage at the Zoo," and "Cage Service," zoos attended to the comfort, health, and diet of zoo animals individually and with great care. Indeed, it may have seemed that the animals in the exclusive zoo community dined from a more interesting menu than zoo visitors. The daily meals of a prized baby gorilla included "two quarts of milk mixed with one raw egg, a few slices of bread and honey, zwieback, five bananas, an orange, prunes, grapes, and half a head of lettuce." A bird house required "a wider variety of foodstuffs than the Waldorf-Astoria." According to another article, "The zoo is the strangest hotel in any city. Unique guests comprise its registration list—guests who come from many lands, speak a thousand languages and whose personal requirements and appetites vary almost to the infinite." Such articles portrayed zoo residents as an elite, pampered class of animal celebrities, and reported on their habits with the fervor of gossip columnists. The zoo was an exotic and colorful neighborhood of the city, and the daily lives of its inhabitants were easy for newspaper readers to relate to—births, deaths, diets, and so on.[35]

People who gave up their pets to the zoo thus felt justified in requesting special treatment for them. A Japanese robin, for example, liked "his mirrors very much" and was fussy about how his food was served: "Mockingbird food in one compartment of tray, and crushed hemp seed in the other. Two or three meal worms daily in morning, sometimes only one or two." The fact that "so many persons bring in pets with all kinds of restrictions attached to them and all sorts of instructions as to their care" dismayed Belle Benchley, director of the San Diego Zoo. Still, she understood and seemed to agree with the motivation to donate pets. At the zoo, a family's unwanted pet would "have better care than they could give it."[36]

People who donated their pets saw the zoo as lodging for the privileged animal members of a larger human-animal community. Whether pets were left temporarily or permanently, they became part of an animal family, as well as pets to all zoo visitors. The zoo was perceived as a place for exceptional animals rather than type specimens; people detailed the accomplishments and good manners of the pets they wished to donate. Parting with a pet was often difficult. People who wrote to the National Zoo to inquire about donating their pets sought assurance that they had made an ethical decision. "Do you want an extra large all white cat for the zoo?" asked Miss E. Fish, of Washington, D.C., "And would you treat him kindly?"[37]

Elephants for Civic Unity

Donating animals to the zoo, however, was not just a matter of individual initiative. It could also be a community project. In 1914, the Boston *Post* coordinated a campaign for the city's children to donate their pennies to purchase three retired vaudeville elephants for the new Franklin Park Zoo. The *Post*'s campaign drew support from tens of thousands of people across the spectrum of class and ethnicity. The newspaper claimed that seventy-five thousand individuals contributed to the elephant fund, which raised $6,700 over three months. In coordinating this community effort, the *Post* played a role similar to that of reformers and patriotic societies who sought to boost civic solidarity and to revitalize community identity through public displays such as pageants, parades, and "safe and sane" Fourth of July celebrations. As the motto of one such pageant stated it, "If we play together, we will work together." The *Post*'s elephant campaign sought to foster a spirit of teamwork by drawing together a diverse city for a cause that everyone could support: a project for children. The zoo thus became an emblem of citywide cooperation and fellowship.[38]

The Boston elephant campaign was not unique. Children's campaigns purchased the elephants Clio, in 1890, and Maude, in 1902, for the Atlanta Zoo. In the early 1920s the children of Baltimore contributed their pennies to buy the elephant Mary-Ann. In St. Louis, the city's children purchased an elephant with their pennies and named it Jim, after the president of the board of education. The newspaper the *New Orleans Item* coordinated school children in a movement to obtain an elephant for that city's zoo in 1923. In Indiana, the *Evansville Courier* sponsored an elephant fund in the 1920s. Such campaigns followed a tradition of newspapers raising money for Christmas seals, or to help sick children.[39]

During the months of fundraising and anticipation in Boston, numerous meanings were written onto the elephants. As cultured, trained animals, they held little status as natural objects. In the eyes of their vaudeville owners, they were pets. The *Post* cast the elephants as instruments of moral instruction for children, and as symbols of childhood innocence for adults. The elephants provided the newspaper with a vehicle for conveying community spirit to its readers. On the vaudeville stage, they had been entertainers and world travelers—outsiders without strong ties to the city. Throughout the campaign the elephants proved themselves to be models of virtuous citizenship. On entering the zoo they relinquished the rootless life of the theater and became permanent residents of Boston. Public support across boundaries of class and ethnicity focused on assimilating the elephants into the zoo and bolstering the zoo as a symbol of civic virtue.

Pennies for Elephants

The Franklin Park Zoo in Boston opened officially on October 14, 1913. On that day a crowd of five thousand gathered to see three new permanent buildings—an "oriental" bird house, an open-air flying cage for birds, and bear dens. The zoo had been a long time in planning. The Boston Natural History Society had first proposed exhibits of local fauna in 1887. But the city parks department demanded that the Society build and manage such a display, and the Society was unable to raise the funds to do so. In 1910, a new park commissioner—architect Robert Peabody—decided that Franklin Park was underused. An influx of private money made it possible to build an "attraction"; Peabody decided on a zoo. He consulted William T. Hornaday, director of the successful new zoo in the Bronx, on choosing a site in the park. Hornaday convinced Peabody that the zoo needed exotic animals as well as American ones in order to attract a crowd that would justify the cost of its maintenance. Within a few months of opening, the zoo had on display twenty bears of various species, three elk, a deer, a llama, ocelots, wolves, raccoons, a hyena, a leopard, a lynx, and a monkey. The zoo's "first large animal of any consequence" was a buffalo named Bill, reported—with appropriate skepticism—to be the offspring of the model for the buffalo nickel. It was a modest collection, the result, mainly, of the efforts of park planners.[40]

Six months later, William Orford, owner of three elephants trained for vaudeville, offered to sell his performing troupe to the city of Boston. Like the people who gave their pets to the National Zoo, Orford perceived the Franklin Park Zoo as a retirement home for his elephants, which he described as pets. Turning down offers as high as $15,000 from circuses, Orford set his price at $6,000, on the condition that the city would present the elephants to the zoo, where they would have "a nice home and good treatment for the rest of their very long lives." Orford and his wife, who performed with the elephants, themselves wished to retire, in Germany. The trouble and expense of transporting their pets home were too much.[41]

Editors of the Boston *Post* sensed an opportunity to both boost readership and serve the community: on March 9, 1914, the newspaper ran a front-page story with the banner headline "How the Children Can

Secure Trained Elephants for City Zoo," leading off a campaign for children to contribute their pennies and other small change to buy the elephants. The newly appointed "elephant editor" proclaimed, "Everyone agrees that a zoo without an elephant is not a real zoo at all." For the next three months the newspaper printed stories about the fundraising campaign nearly every day. Every donation, from one cent to hundreds of dollars, was acknowledged in a daily list of contributors. To heighten interest in the promotion, the *Post* printed photographs of the elephants, which were still performing in local theaters, as well as the letters and photographs of children who donated money.[42]

On the first day of the fundraising campaign, the *Post* obtained endorsements from an array of civic leaders, beginning with the governor and the mayor. Members of the school committee testified to the educational importance of a zoological collection. A veterinarian certified the animals' health. The chairman of the park commission reiterated that the zoo would be incomplete without elephants. And the pastor of the Deadly Street Baptist Church praised the effort as "a commendable thing." A representative of the Massachusetts Society for the Prevention of Cruelty to Animals had to agree that the animals would be better off retired from their careers as traveling performers. Along with his $5 contribution, he wrote, "I know the hardships of the transportation of animals even under the best of conditions, and I feel they would be much better off in a permanent home here." If anyone opposed elephants in the zoo, their voices were not heard in the *Post*. In any case, it would have been churlish to criticize "so delightful a cause." The elephants would benefit the city's children. As the innocent symbols of Boston's future, children provided a cause behind which diverse groups could rally together.[43]

The *Post* inaugurated the campaign with a contribution of $500 toward the $6,000 needed to buy the elephants. Prominent citizens quickly rose to the challenge, and local businessmen pitched in. Within a day the fund was up to $1,379, and the *Post* printed the first contributors' names on the front page. "The idea is to give the fund a good send-off with a few sizable subscriptions from friends of the children," wrote the elephant editor.[44]

The *Post* appealed to reader sentimentality and nostalgia for lost youth by printing letters from the children who donated their "mites."

"Here are some more of those little folk letters," announced the elephant editor. "Don't fail to read every one of them, for many sweet memories of years ago are to be found between the lines." While tugging at misty-eyed readers' heart strings, the letters were also meant to loosen their purse strings. "Those who love children, whose own hearts beat more warmly as they watch little figures pressed close to the railing, and see eyes grow large with wonder at the sight of the great mouse-colored elephants, will give freely of what they have."[45]

In response, the clerks of a division of the Boston Post Office donated $7. The Irish-American Hockey Team offered to play a benefit game. The Park Riding School made its facilities available for a benefit "society horse show." Members of a jury sitting on a Civil Court case passed the hat during a recess. The Ancient and Honorable Artillery Company collected $64.62 at its smoker. The City of Boston Assessors Department chipped in $20.75; the Penal Department gave $4. And passengers on the 7:42 commuter train from Sharon collected precisely $7.42. All received public acknowledgment and praise in the newspaper.[46]

Again and again, however, the elephant editor discouraged grown-ups from contributing large sums and pressed for small donations from children. "The bulk of the $6000 needed shall come from the children themselves in order that they may have a direct personal interest in these fascinating animals. . . . When each child feels a sense of personal ownership, how much more attractive [the elephants] will become." Few children, presumably, read the newspaper or had much small change to contribute. The elephant campaign thus became a way for the newspaper to guide parents in teaching their children a moral lesson about the value—and fun—of hard work and of working for a community cause. To earn pennies, nickels, and dimes for the elephants, children gave up candy and movies, went to bed without complaint, and did chores and errands. The editor also suggested that children raise money by holding "elephant parties," to which they would charge their friends admission and play pin the trunk on the elephant and other games adapted to the elephant theme. The newspaper published a cartoon depicting the "magnificent battle" to raise the money, with a sun of "grown-up encouragement" shining down on a parade of children, led by the elephant editor, marching up Elephant Campaign Hill toward the Franklin Park Zoo. The *Post* also stated explicitly how children's activities would promote community among adults: an elephant party could be "kind of a visiting day for the mothers of the neighborhood."[47]

The newspaper promised children and parents rewards for their efforts. In the short term they could enjoy the novelty of seeing their names printed in the paper. The *Post* reminded its readers, "No sum is too little. . . . If your name or your little boy's or little girl's name has not yet been printed, hurry and send it in." The elephant fund also represented a long-term investment. A campaign for the benefit of children signified a promise of civic unity for the future. And the *Post* assured its readers that elephants would be present in that future as reminders of the community effort to purchase them. Elephants were so long lived, the *Post* claimed, that not only would readers' children enjoy visiting them in the zoo, but so would their children's children.[48]

The elephant editor provided further evidence that children from diverse backgrounds could work together by printing children's letters that indicated a wide range of social and economic circumstances. Benton Bradshaw's mother gave him a penny for every rat he caught—he earned ten cents. By contrast, a six-year-old girl earned twenty-five cents by bringing home "stars and hundreds" from school. Sarah Wilson "earned 5 cts. doing some sewing for a lady." Children of Boston's various ethnic groups—Mary Flynn, Josephine Costello, and Arthur G. Schwarzenberg, for example—each sent their contribution. In addition, the *Post* printed photographs of child contributors that reinforced the message of a diverse community, both common folk and high society, united behind a common cause. The March 28 edition printed a snapshot of "Stanley Nichols, of Burlington, Mass.," who sold eggs to earn money, dressed in overalls. Next to him was a portrait of Evelyn V. Sullivan of Roxbury, posed in a frilly dress, who "after dreaming of Tony [one of the elephants] sent 25 cents to the fund."[49]

Making the Meaning of Zoo Elephants

The *Post* also used the elephants themselves to further its agenda of building community spirit and teaching children good behavior. As animals removed from nature, and with long associations with human culture, the elephants were particularly easy symbols for the *Post* to manipulate. To zoo audiences at the turn of the century the natural habitats of the elephant were the circus ring and the vaudeville stage. The Boston public already knew Orford's elephants—the two adult females Mollie and Waddy, and the "baby" male Tony—as clever performers at Keith's Theater. Zoogoers perceived elephants as practically domesticated. To be sure, elephants came from "wild country" in Asia and Africa. But performing elephants, more immediately, came mainly from India. And in India, they only served different masters from those in the United States, as the property of royalty or as working animals.[50]

In fact, the degree to which the *Post* ignored or misrepresented the status of the elephants as natural objects is striking. In the first announcement that the elephants had been offered to the city, they were described as "Burmese" elephants, aged sixty, forty-three, and three years. Later, the newspaper reported that the adult elephants had "been secured from the herd of one of the ruling princes of India," and were both thirty-four years old, and the baby six. In another article, confusing their species, the elephants were described as "three African pets." The newspaper occasionally offered further tidbits of misinformation, such as "Elephants usually get their full size when about 30 years old, and during the remainder of their 200 or more years of life they don't grow." A series of articles under the heading "Little Elephant Stories" started out with an attempt at natural history description, but quickly turned to tales of Mollie, Waddy, and Tony's antics while on the road as a vaudeville act. The *Post* portrayed the elephants as having a place in human culture rather than in nature.[51]

The *Post* also played to adults' nostalgia by publishing stories about earlier zoo and circus elephants that had become local and national pets. Everyone knew the story of Jumbo, the London Zoo elephant purchased by P. T. Barnum and brought to the United States in 1882. "[W]hile the boys and girls of today perhaps never even heard of Jumbo, I am sure that if they ask their fathers and mothers about him they will learn a great deal," explained the elephant editor. One child wrote to the *Post*, "I am a little English boy, and my mother often tells me how she used to ride on Jumbo's back in the zoo in London." By evoking a tradition of elephants as community pets, the *Post*'s three-month campaign conveyed the value of zoo elephants to the city's social solidarity.[52]

In addition to fostering the idea of elephants as pets, the *Post* used the elephants as instruments of moral instruction in "Little Elephant Stories," which taught lessons about how children should behave by drawing parallels between the experiences of trained animals and children. According to the elephant editor, "the elephant is about the best example for a boy and girl we know of. Above all else they are very obedient. When they are directed to do a thing they do it at once." Mollie and Waddy, the older female elephants, assumed the role of kindly aunts or of model older sisters, "not a bit jealous" when Tony got extra attention. A photograph showed them sitting on barrels having a tea party—part of their act—with the caption: "Mollie and Waddy handle their cup and saucer very nicely indeed and never, never forget their manners." Tony, the baby, was simply not capable of such adult standards of behavior. Instead, he assumed the roles of mischief-maker and underdog; his owners looked on Tony's roguish behavior with a boys-will-be-boys indulgence. Children seemed to identify with Tony; he was the elephant mentioned most frequently in their letters. Tony got in trouble for stealing sugar and pastries, but redeemed himself by good behavior later. He even picked up a dime off the street and donated it to the elephant fund. If Tony was "a little slower in his movements than the big ones," he made it up by playing jokes on them. Whenever a photograph showed only the adult animals, the caption explained that Tony, "the little rascal," was hiding.[53]

The stories were fiction, although Tony's real-life behavior may have inspired the impish disposition that the *Post* writer drew for him. Circuses often foisted their unmanageable elephants on zoos, which presented them to the public as gentle giants. An article about Mollie, Waddy, and Tony published twenty-five years after the *Post*'s campaign referred to them as "three trick elephants which were being disposed of because of their bad tempers." At the National Zoo, the first animals in the collection were elephants named Dunk, who had a reputation for charging at other elephants, and Gold Dust, who was considered "mean, treacherous, and a man-killer."[54]

In any event, the *Post* further described the elephants as docile, gentle, and sensitive. According to the newspaper, they had been trained only with rewards of patting, praise, and sugar, and had never injured anyone. In short, they were perfect companions for children. Indeed, the reporting about the elephants resembled a regular feature of the

Post—a series of animal profiles that ran under the headline, "Now Here's the Story of a Very Clever Pet." The transition from their working life to the zoo brought the elephants closer to being considered members of a human-animal family. Other pets—cats, dogs, and birds, for example—already belonged to this extended social group. Many of the children who wrote letters to the *Post* made donations in the names of their pets: a "Scotch Collie" named Rexall A. Hobbs, a canary named Reginald (who submitted a poem, as well), Nemo (another dog), and Dixie the cat, among others. "I hope the Elephants may have as nice a home as I have," wrote Bunny, a Roxbury Kitten. The elephants thus joined a community of pets. At the zoo they would play with children, giving them rides. Children looked forward to feeding the elephants peanuts and other treats.[55]

The elephant campaign culminated in a community-building event that celebrated virtue, nationalism, and popular entertainment. On June 7, 1914, the *Post*'s front page headline read: "50,000 Children Shriek Welcome to Elephants." The audience had overflowed the thirty-three-thousand-capacity Fenway Park, a bigger crowd than either the World Series or Harvard vs. Yale football games attracted. The elephants provided the climax to a two-hour program that began with band music and continued with dancers, acrobats, clowns, and an impersonator of Teddy Roosevelt. At the end of the festivities, the elephants were presented to the city, to be maintained in the Franklin Park Zoo, where they would "not be used for commercial stage exhibitions." For a grand finale the elephants waved American flags with their trunks while the band played "America." It was their last performance. At the zoo, the elephants were no longer celebrities of the stage—they became "regular Bostonians," patriotic citizens of the city. They had been portrayed as models of integrity, setting a standard to which the city's residents could aspire. If the children of Boston were as well behaved as the elephants, the elephant editor implied, then the city's harmonious future would be secure.[56]

Collections and Souvenirs

The stories surrounding animal donations from the public help explain why an elephant was indispensable to a representative zoo collection. A zoo was not a zoo without an elephant in part because the animal

MORNING, JUNE 7, 1914 **   PRICE, FIVE CENTS

50,000 CHILDREN SHRIEK WELCOME TO ELEPHANTS

Fenway Park Packed With Cheering Small Humanity— Greatest Crowd in Inclosure in America's History—All Seats Taken Early and Overflow Takes to Field— Thousands Unable to Gain Entrance to Park—Governor Presents Mollie, Waddy and Tony to City and Mayor Accepts—Stirring Scenes on the Children's Gala Day—No Accident—Elephants in New Home at Franklin Park

FIGURE 9. The presentation of three vaudeville elephants to the city of Boston was front-page news. Boston *Post,* June 7, 1914, p. 1.

pleased everyone—it satisfied diverse expectations for who belonged. While there was some negotiation about which animals could represent the small, the fierce, the flying, or the furred members of the animal world, there was no substituting something large for the elephant. By virtue of its size, the elephant—the largest land animal—was a bookend to a representative natural historical collection. At the same time it was almost freakish—it provided the entertainment value of the extreme. While not common, an elephant also was obtainable, unlike gorillas, for example, which were extremely rare and short lived in early-twentieth-century zoos. And the acquisition of an elephant represented a substantial commitment of resources for feeding, housing, and caring for the animal (although these could be offset by selling rides on the elephant's back). A zoo with an elephant was a permanent institution, and the elephant was an awe-inspiring attraction, a model citizen, and a "handsome municipal ornament." It was big enough to represent all the animals in the zoo and to make that institution an emblem of civic pride, a symbol of a city's human population and its potential.[57]

In addition, in the process of their removal from one context to another, pets, elephants, and other animals acquired the status of souvenirs. The pageant and performances of the elephant day celebration in Fenway Park made the elephants into mementos of those events. A souvenir derives its value from the participation of its owner in some special activity or event. The *Post* had encouraged Boston's children to take possession of the elephants by donating their pennies. A newspaper clipping with a child's name on it, pasted into a scrapbook, could facilitate this feeling of ownership. If the *Post* encouraged children to exercise their dominion over the natural world by helping to buy elephants for the zoo, this was a gentle, personal, and perhaps well-meaning display of proprietorship.[58]

Just as important as contributing pennies was taking part in the event that brought the elephants to the zoo. It was a day never to be repeated, and membership in the group that witnessed it boosted the participants' perception of ownership of the elephants. The animals themselves could then serve as souvenirs, touchstones for remembering the elephant campaign, the feelings of fellowship it generated in the city, and the exciting day on which the elephants came to the zoo.

They remained in the zoo as traces of that experience. Mollie, Waddy, and Tony also offered zoo visitors a different kind of transcendence—nostalgia. The elephants evoked memories and stories of other elephants, like Jumbo, and more generally nostalgia for the innocence and capacity for awe of childhood. Through such associations animals came to represent much more than zoological specimens.[59]

Zoo animals were both individual personalities and members of a collection. Pets who were put on display and elephants making the transition from vaudeville to the zoo became members of a privileged community. Ensconced in their orderly houses and paddocks, having often met standards of virtue in order to gain zoo residence, the animals became naturalized, residents of a park who were stripped of the cultural associations of their previous lives. They made up a well-behaved city of beasts within the larger city, a community that required the stewardship of local citizens.

It would be easy to dismiss people who donated animals to zoos as marginal to the project of procuring and displaying a collection of wildlife in a zoo. They were amateurs, hucksters, sentimental pet lovers, and children. As a demographic group, they are difficult to characterize in detail. And often the animals they offered never became part of the zoo collection. But the record of their activities provides an opportunity to examine their ideas about who belonged in the zoo and their concepts of relationships between urban humans and wild animals. Although they do not represent the entire zoo-going public, the active involvement of these people in making zoos reveals more complex interpretations of the zoo than simply the opportunities for education and recreation touted by zoo planners. Similar negotiations between zoos and their publics have continued for a century, and the diverse ways in which visitors interact with animals have both frustrated the attempts of zoo administrators to convey programmatic educational messages and contributed to the long-lasting appeal of their institutions.[60]

On July 13, 1922, an animal dealer named Ellis Joseph delivered a duck-billed platypus to the Bronx Zoo, along with a large collection of kangaroos, birds, lizards, and snakes. The platypus was the first to be seen alive outside Australia, the lone survivor of five animals that began the journey by steamer to San Francisco and then by rail to New York. Joseph had collaborated with an Australian naturalist to build a portable den in which the animals could travel. This unique cage consisted of seven compartments, rising in steps and connected by a ramp. The bottom chamber contained water for a swimming pool, to simulate the environment in which the platypus spends most of its time. The top chamber was a dry sleeping den. Between the intervening compartments, Joseph set rubber squeegees, so that the platypus squeezed excess water out of its fur while walking up the ramp to its den. A diet of earthworms sustained the platypus during its trip; at the zoo it was also fed small shrimp. The Bronx Zoo cautiously put this "most strange and wonderful of all land animals" on display for just one hour per day. It lived only forty-nine days at the zoo.[1]

Zoos depended on animal collectors and dealers like Ellis Joseph. Although some animals came to zoological parks through donations, trades, or loans, zoos could not have come into existence, and their collections could not have taken the shape that they did, without a commercial trade in wild animals that provided a reliable supply of polar bears in San Antonio, for example, prairie dogs in Philadelphia,

FIGURE 10. The Australian animal dealer Ellis Joseph was well liked by zoo people. © Wildlife Conservation Society, headquartered at the Bronx Zoo.

FIGURE 11. A zookeeper shows a platypus to visitors at the Bronx Zoo in
1922. © Wildlife Conservation Society, headquartered at the Bronx Zoo.

or zebra in Denver. Without the animal trade few zoos could have displayed more than deer and birds—they never would have become
"real" zoos with elephants.

The animal-dealing business was well established by the late nineteenth century, both internationally and within the United States. It
had started in Europe, where dealers dispatched collectors to areas of
Asia and Africa under colonial rule. Some of the animals captured
made their way to American zoos. In the United States, during the second half of the nineteenth century, the trade in North American and
other wild animals became organized, evolving from a sideline of pet
stores and collectors of dead specimens into a lucrative business in its
own right. New zoos in the United States, as well as in Europe, British
colonies, China, Japan, and South America, helped fuel the growth of
the wild animal business. But zoos constituted only part of the market.
Animal dealers also sold to circuses that were expanding their

menageries and trained animal acts, private fanciers of animals and birds, the pet trade, and laboratories.[2]

Few zoos could afford to mount their own animal collecting expeditions. The collections that zoo directors assembled for the education and entertainment of their audience thus reflected what was for sale rather than representative animals from across the taxonomic order, or from every continent, country, or geographic region. For their part, collectors and dealers had different priorities from zoo directors. They worked within the parameters of the business world. Like other businessmen, they had to make decisions about investments of time and money. They tended, for example, to deal in animals they were certain to sell—brightly plumed birds in demand by private fanciers, or elephants, tigers, and camels for circuses. The demand of circuses for big cats and elephants made such animals readily available to zoos. In addition, dealers in live specimens focused on animals that adapted well to life in captivity and were good investments—likely to survive months, and sometimes years, of travel under harsh conditions. But a rare specimen—a platypus, for example—could bring prestige to both an animal collector and the zoo the animal was sold to, and it was worth investing time, effort, and money in learning how to care for it. Thus the market for wild animals, and the abilities both of animals to acclimatize and of collectors to care for them, shaped the collection of animals on display in zoos.[3]

The heyday of the animal trade spanned the decades between the careers of the German Carl Hagenbeck in the 1860s and the American Frank Buck, who worked into the 1940s. Examining this business reveals more than the logistics of how exotic animals traveled to American cities. During these years, collectors and dealers in live wild animals rose in social status from obscure, marginal figures to heroes of popular culture. Animal collecting developed an identity as an occupation. The behind-the-scenes work of providing animals to zoos was made visible to the zoogoing public through newspaper and magazine articles, as well as books, that celebrated the trials and triumphs of animal collecting. In these accounts collectors gave an insider's perspective on their business, which required straddling the worlds of colonial commerce, circus entertainment, and zoo science. They sometimes contrasted their work with hunting, which had different goals and ethics. Their adventure stories added meaning to the specimens they delivered to zoos by making the animals into living links to places, people, and experiences described in their books. For zoo visitors familiar with the story, the rhinoceros in the Bronx Zoo was not just a rare creature, but the animal captured by the Nepalese General Shum Shere and delivered by Frank Buck, who recounted the story in *Bring 'Em Back Alive*. In addition, animal collectors acquired well-honed skills in animal keeping, craft skills that were valuable in the zoo community.[4]

Colonialism and the Animal Trade

Wild animals became one among many natural products of Africa and Asia to be traded in the eighteenth and nineteenth centuries as European nations gained colonial power. The Dutch East India Company, for example, constructed pens and stables at an Amsterdam dock and maintained a menagerie at the Cape of Good Hope. Naturalists traveling abroad returned home with live specimens, and animals arrived in European ports as the souvenirs of sailors and merchants. A few exotic animals were also imported to the United States—a "Lyon of Barbary" in 1716, a camel in 1721, a polar bear in 1733, and an elephant in 1796. Until the mid-nineteenth century, however, American menagerie owners generally traveled to London to buy animals.[5]

The Zoological Society of London, organized by colonial administrator and naturalist Sir Stamford Raffles, took advantage of the influx of wildlife that accompanied naturalists and travelers. Its gardens opened in 1828. Unlike the elite Zoological Society, however, the commercial animal trade had a seedy reputation to match the docks of the port cities where it was based. "Every one has seen something of the business of a dealer in animals in its primitive form," wrote the secretary of the Zoological Society of London, in 1909. "Near the London Docks, and on the quays of great shipping ports like Havre and Marseilles, there are to be found untidy and generally evil smelling little shops crowded with parrots and monkeys, and similar casual acquisitions from sailors. The proprietors . . . find out what creatures they can sell most readily, and give orders to sailors or petty officers, sometimes on speculation and sometimes at the request of customers."[6] By the mid-nineteenth century European businessmen who had built up a sideline in trading exotic animals on the local docks began turning to animal dealing as their main source of income. The port city of Hamburg developed into an important center for the live animal trade. There Carl Hagenbeck became the best known of German animal dealers, and the leader in moving animal dealing from the margins to the mainstream of colonial trade.

Hagenbeck turned animal dealing from an erratic, dockside affair into a systematic approach to collecting and distributing animals, and he supplied thousands of animals to American zoos. By the turn of the twentieth century, the name Hagenbeck—promoted through traveling animal shows and popular articles and books—was synonymous with "animal collector" around the world. Hagenbeck's career provided a model for others to follow; the American dealer Frank Buck, for example, considered himself Hagenbeck's successor.

The Hagenbeck company traces its origins to Carl Hagenbeck's father, a Hamburg fishmonger. In 1848, sturgeon fishermen brought him six seals that had become tangled in their nets. The elder Hagenbeck put the seals on display, charging eight pfennig to curious onlookers. When local interest in the seals waned, he took them to Berlin, where—in the midst of violent political unrest—he still managed to turn a profit with the exhibit. Following this success Hagenbeck began buying animals on the Hamburg docks to keep in his private menagerie, including, in 1852, a full-grown polar bear. In 1859, before Carl's fifteenth birthday, and at the end of his formal schooling, his father presented him with a choice: either to continue in the successful Hagenbeck fish business, or to take over the animal collection as a separate enterprise. Carl Hagenbeck became an animal dealer.[7]

Carl Hagenbeck continued to meet ships at the Hamburg docks, and he attended the annual fall auctions of animals at the Zoological Gardens in Antwerp. Like other animal dealers, to this point he was a broker for animals that had already arrived in Europe. Hagenbeck, however, sought a more reliable line of supply. In 1864, he met an Italian explorer named Lorenzo Cassanova, who had collected animals during his travels to northern Africa. Cassanova became the first "traveler" that Hagenbeck contracted with to collect animals. Cassanova made two expeditions for Hagenbeck. In 1870, Hagenbeck met the sec-

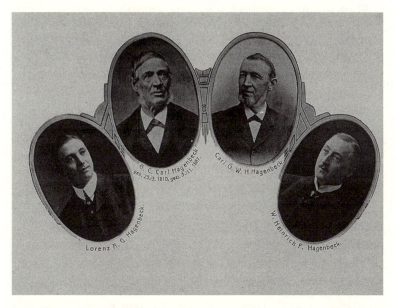

FIGURE 12. Three generations of the Hagenbeck family participated in the animal dealing business. The famous Carl Hagenbeck is third from left. Also shown are his father, G. C. Carl Hagenbeck, and his sons Lorenz and Heinrich. From a souvenir book titled *Carl Hagenbecks Tierpark Stellingen, Ein Ruckblick.*

ond one in Suez, at the Suez Hotel. Cassanova, feverish and on his deathbed, had sent for Hagenbeck to meet him and take charge of his caravan of captured animals. "I shall never forget the sight which the courtyard presented," Hagenbeck wrote. "Elephants, giraffes, antelopes and buffalo were tethered to the palms; sixteen great ostriches were strolling about loose; and in addition there were no fewer than sixty large cages containing a rhinoceros, lions, panthers, cheetahs, hyenas, jackals, civets, caracals, monkeys, and many kinds of birds."[8]

Loaded down with hay, bread, and other food, and leading one hundred milk goats for nursing the baby animals, Hagenbeck journeyed on with the menagerie to Alexandria, but not without mishap. "On the way to the station the ostriches escaped," he reported. "Then one of the railway trucks caught fire." In Alexandria, Hagenbeck joined forces with another contracted collector, named Migoletti. It was the biggest shipment of African wildlife of Hagenbeck's career. The animals were taken by boat to Trieste, and then over land to Vienna, Dresden, and

Berlin. At each stop Hagenbeck sold some animals to zoological gardens. The remainder went to Hagenbeck's headquarters near Hamburg. This extraordinary shipment established Hagenbeck's reputation as an animal dealer; stories about the adventure circulated in the popular press and were retold by Hagenbeck himself in his autobiography.[9]

Hagenbeck expanded his business further by nurturing commercial, political, and military contacts in Germany's African colonies. He supplied camels to the military stationed in German Southwest Africa. Another animal dealer observed that "Every German consul in the smallest port of Asia or Africa had been a Hagenbeck agent on the side." Hagenbeck also dispatched animal collectors around the world. He sent one expedition, for example, to Mongolia, to catch specimens of the wild Prjwalsky's horse. The market for his exotic animals in zoos, circuses, and traveling menageries was so large that, he wrote, "it was sometimes impossible for me to satisfy all the demands which I received." At the height of his career Hagenbeck employed fifty animal catchers and operated five holding stations for animals in Asia, several in Europe, and two in the United States. Between 1866 and 1886, he imported, among other animals, over a thousand lions, three hundred camels, one hundred and fifty giraffes, tens of thousands of monkeys, thousands of crocodiles, boas, and pythons, and more than a hundred thousand birds. Hagenbeck's profits rivaled those of importers of sugar, coffee, and palm oil.[10]

Management of the business remained in the family. Hagenbeck's sister Christiane oversaw the birds. His stepbrother John operated a holding station in Ceylon. His brother Wilhelm trained and helped manage the animals at home; and his brother Dietrich (the only Hagenbeck to attempt collecting in the field) died of black water fever in Zanzibar in 1873 while on an expedition to collect hippopotamuses. The Hagenbeck animal dealing business spawned a circus, traveling ethnographic exhibits, and finally a Tierpark, or zoo, which opened in 1907. After Carl Hagenbeck died in 1913, his sons carried on the business.

American zoo directors visited Hagenbeck when they went to Europe. Those who did not travel, however, could buy animals through Hagenbeck's American agents. The Cincinnati Zoo's secretary, Lee Williams, was Hagenbeck's American broker for several years, and in

1902, the zoo's director, Sol Stephan, took over this job. Animals on display in Cincinnati were sometimes Hagenbeck's stock en route to or from the Hamburg headquarters.[11]

Although Hagenbeck operated the largest animal dealing business in Europe, it was not the only one. In London, Charles Jamrach dominated the animal trade; one of his local rivals was William Cross. German dealers were more successful than British ones, however, in expanding their business to the United States. Around 1844, for example, the German brothers Charles and Henry Reiche emigrated to New York; by 1853 they had imported twenty thousand canaries (eight thousand went to San Francisco gold miners). Circus owners—in particular Isaac Van Amburgh, the first notable American wild-animal trainer—asked the Reiches to import other animals. About 1860, Charles Reiche returned to Germany and set up a business based near Hanover that was similar to Hagenbeck's, although on a smaller scale. The Reiches recruited animal collectors, maintained agents in Egypt and Ceylon, and shipped their stock to Hamburg, before it went on to the United States or customers in Europe. After the deaths of Charles Reiche in 1885 and Henry Reiche in 1887, the business passed to Paul Ruhe, who had been one of their animal collectors, and who was presumably a relative of Louis Ruhe, head of an animal dealing business with offices in New York, Hanover, London, and New Orleans. German connections to the American exotic animal trade continued into the twentieth century with Henry Bartels and Henry Trefflich, German immigrants who imported animals into New York through Germany. Julius Mohr and Arthur Foehl were additional German animal dealers at the turn of the century who maintained offices in the United States. It was through animal dealers based in Germany that most wildlife from Africa, South Asia, and the East Indies entered the United States.[12]

The Market for Wild Animals

To a large extent, the success of Hagenbeck and other animal dealers paralleled the boom in circuses in the late nineteenth and early twentieth centuries. Between 1880 and 1930 the number of American zoos grew from about four to more than one hundred. During the same period more than 650 different circuses, Wild West shows, and dog and

FIGURE 13. Circus shows boosted demand for elephants from animal dealers and made the animals more readily available to zoos. Reproduced courtesy of Feld Entertainment, Inc.

pony shows performed at one time or another in the United States, reaching their peak in numbers and popularity around 1920. Wild animals became increasingly important components of the performance. One-upmanship among circus owners increased the demand for elephants, camels, bears, and wild cats. P. T. Barnum first visited Hagenbeck in Germany to buy animals in 1872. In the 1880s, Hagenbeck filled huge orders for elephants from Barnum and his competitor Adam Forepaugh. In 1875 each circus had four elephants; by 1883 Forepaugh owned twenty-five elephants. A combined Forepaugh-Barnum show in New York City's Madison Square Garden in 1887 featured sixty elephants. In 1883 alone, Hagenbeck had exported sixty-seven elephants from Ceylon.[13]

After the turn of the century, wild-animal acts, in which a trainer—sometimes armed with chair and whip—entered the ring with an assortment of "Bloodthirsty Savage Beasts," became increasingly common. Particularly popular were acts like Alfred Court's, where a mixed group of animals, "Natural Enemies Since the Dawn of Creation," posed in a peaceable tableau. Court worked the most elaborate animal group in the business—lions, tigers, black jaguars, snow leopards, black panthers, pumas, cougars, Great Dane dogs, polar bears, Himalayan bears, spotted leopards, mountain lions, spotted jaguars, ocelots, and black leopards. Another trainer, Clyde Beatty, achieved fame in the 1920s for facing off against lions and tigers in the circus arena.[14]

Animal dealers tailored catalogs to this market, advertising trained groups of animals for sale or lease, and distributed the leaflets to both zoos and circuses. In 1913, Hagenbeck offered for $4,000, "1 Group consisting of 5 Polar Bears and 2 Brown Bears, about 1 year old, 3 Black Tibet Bears, about 1 1/2 years old, now being trained together," as well as "1 Group of trained animals: 1 Female Indian Elephant, 'Jenny,' 7 feet high, 2 white camels, stallion and gelding, 1 Piebald Pony stallion, 1 Collie Dog, 1 Rhesus Monkey, the lot," for $4,250. The American dealer I. S. Horne in 1913 had for sale "1 Male African Lion, 3 years old; 2 African Lionesses, 2 1/2 years old; 2 male Pumas, 2 years old; 2 Brown Bears, 2 years old; arena, all props, shipping cages, etc. complete" for $5,000.[15]

Dealers coordinated the steady supply of animals in demand by circuses. Since zoos bought from the same dealers as circuses, it was easy for them to obtain the species of animals shown in circuses. Big cats sold for prices ranging from $90 to $1,600 depending on the age, health, and rarity of the specimen. The varieties that performed in the circus ring were the same as those found in zoo carnivore houses. Both one-humped and two-humped camels were always available, at about $300 each. Other animals typically exhibited by zoos were also staples of the circus menagerie. The animals featured in the 1928 Sparks Circus parade were representative. In cages pulled by "dapple gray horses," the parade featured monkeys, polar bears, lions, spotted deer, tigers, leopards, kangaroos, ostriches, hyenas, black leopards, and sea lions. Bringing up the rear were "Five camels in single file with grooms dressed in Turkish costumes," three zebras, three llamas, and "Nine elephants in single file, a mahout on the head of each, the lead bull carrying a girl in a howdah." Minus their costumed keepers, any of these animals could have been displayed in the zoo.[16]

The boom in the elephant trade made more elephants available to zoos; in fact, there was a great deal of exchange of animals among circuses and zoos. Some circus elephants eventually became zoo

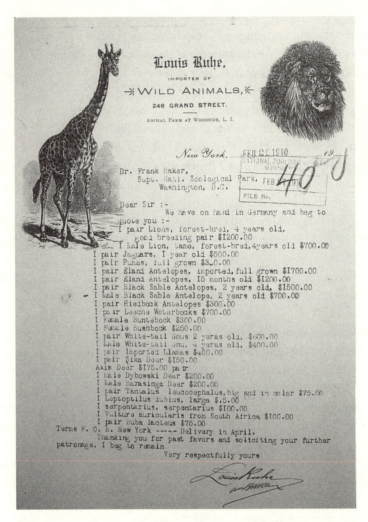

FIGURE 14. A 1910 price list from animal dealer Louis Ruhe. Smithsonian Institution Archives. Record Unit 74. National Zoological Park, 1887-1966. Records.

elephants, and vice versa. P. T. Barnum, for example, persuaded the London Zoo to sell him the famous and beloved elephant Jumbo in 1882. The Cincinnati Zoo's first elephant had been a circus animal, as had the elephants in the National Zoo. The elephants in Boston's Franklin Park were former vaudeville performers.[17] Some circus animals became zoo animals during their off season, when zoos provided them with winter quarters. This tradition lasted at least through the 1950s. To cite an early example, in 1873, circus owners P. T. Barnum, George F. Bailey, and Isaac Van Amburgh, and animal dealers Louis Ruhe and Charles Reiche, all displayed animals in the Central Park menagerie.[18] Zoos and circuses also traded or sold animals among each other. Lists of animals for sale provided only cryptic descriptions, however, and in 1890 the National Zoo's director was privy to some inside information about animals for sale by the Barnum and Bailey Circus: "The camels are very old and the llamas have a touch of the Mange," wrote William Blackburne, who worked for the circus but was about to become head keeper at the zoo. "The young tigers are weak in the back and each had the misfortune to have half of their tails bit off by the striped Hyena who was in the same cage. . . . Don't mention to Jas. A. Bailey where you get your information."[19]

In addition to an exchange of animals there was much overlap among circus people, animal dealers, and zoo employees. William Blackburne was just one among many circus menagerie workers who tired of life on the road and sought work as zoo keepers. A showman who went by the name of "Colonel E. Daniel Boone" did a lion act with the Forepaugh circus in the 1890s and later performed with other shows. In 1901 he wrote to the National Zoo's director to ask for a job: "I am tired of knocking around. Would like a steady place. When ever you want a head keeper I think I can fit the bill." The Cincinnati Zoo's director Sol Stephan had started out as a circus elephant keeper; later, as zoo director, he doubled as a Hagenbeck agent.[20]

Some dealers, like Hagenbeck, also trained animals for circuses and managed their own shows. Frank Bostock, who imported animals to Baltimore in 1900, is another example. His animal dealership doubled as a private zoo, and he competed with the Baltimore Zoo in Druid Hill Park for an audience. But Bostock was an animal trainer and performer as well; after his Baltimore building burned in 1901, he took his

"Great Zoological Arena" on the road to the Pan-American Exposition in Buffalo and other fairs. Other animal dealers moved back and forth between business and zoo employment. To name two examples: William Conklin, who had managed the Central Park menagerie, joined his wife in an animal importing business in the 1890s; Fletcher Reynolds first worked as an animal collector and then for Hagenbeck agent John T. Benson at his Nashua, New Hampshire, farm. After a stint as director of the Candler Zoo in Atlanta, Reynolds returned to working for Benson in 1935.[21]

A Natural History of Collecting

Although animal collectors and dealers sometimes went to work at zoos, and zoo directors sometimes collected animals, zoo people generally regarded animal dealers with suspicion. Collectors belonged to a different social group from zoo directors. They were businessmen who were affiliated with circuses, which were known for numerous forms of swindling. Zoo people always suspected that collectors and dealers were cheating them. Collecting live animals was a marginal occupation in the late nineteenth century. It was, in the words of a Wichita, Kansas, collector named Charles Payne, "a peculiar work." Nonetheless, Payne added, "very few persons have the faculties required for it. It is comparatively easy to kill game, but capturing it alive is another and much more difficult undertaking."[22] Animal collectors took pride in the skills required for their trade, and in distinguishing themselves from hunters. Some collectors started from an amateur interest in natural history, or from experience in collecting dead specimens for museums. They often specialized in particular species or the fauna of a restricted geographical region that they knew well. Others got interested in collecting through working in circuses or vaudeville. They took advantage of their opportunities to travel overseas to buy and trade animals. During the 1920s and 1930s, many animal collectors became celebrities. The popular books and articles they wrote shed light on how they perceived their work and its special requirements.

In the second half of the nineteenth century, businesses dealing in natural history specimens—minerals, fossils, bird eggs, and taxidermy specimens—sold material to the dozens of new American college and urban museums, as well as to individual natural history enthusiasts. It was the kind of business that could start as a hobby, the outgrowth of a boyhood love of nature. Hunting and camping were considered healthy activities for boys. If they took an interest in collecting as well, books like Daniel C. Beard's *American Boy's Handy Book* provided instruction in a variety of the skills of the naturalist: trapping birds and small mammals, rearing wild birds, collecting and preserving birds' eggs and nests, practical taxidermy, and mounting insects.[23]

By the late nineteenth century, it was possible to earn money as a naturalist, although it was considered "one of the odd ways of making a living," and parents discouraged their sons from turning it into a career.[24] The periodical *The Museum* reported that "The extraordinary process of science . . . have [*sic*] opened hundreds of queer side alleys which lead direct to the avenues of trade." Among them, "a 'rattlesnake farmer,' who lives in the Ozark Mountains, and makes the products of his 'farm' bring money from three different directions. The oil he disposes of to the druggists . . . ; the skins he sells to would-be cowboys, who use them as hat bands, and the skeletons are always a ready sale, the purchaser being the curators of natural history departments of the different college and society museums."[25] On a larger scale, and perhaps the best known of such businesses, was Henry A. Ward's Natural Science Establishment in Rochester, New York. In the 1870s and 1880s, Ward employed collectors and taxidermists (including William T. Hornaday), and supplied museums with finished and mounted geological and zoological specimens and study collections. By acting as a sort of scientific supply house, facilitating both scientific investigation and popularization, Ward became a significant participant in American science.[26]

Although less is known about collectors and dealers in live specimens, many of them seem to have come to their trade and worked in a similar way as collectors of rocks, shells, insects, and stuffed birds and mammals. They advertised in periodicals such as *The Naturalist & Fancier's Review*, which catered to collectors interested in both dead and live specimens. As the result of one such ad, the director of the National Zoo in the 1890s corresponded with W. B. Caraway of Alma, Arkansas, regarding gray wolves, a white pelican, moose, ostrich, and otter, that Caraway was selling. Caraway also offered squirrels, guinea pigs, prairie dogs, an assortment of rabbits, Angora cats, deer, and monkeys, as well as bird eggs and deer horns.[27]

Frank B. Armstrong was a collector based in Brownsville, Texas, who started out in the business of collecting museum specimens. By the 1890s, Armstrong was well known as a taxidermist and collector specializing in the fauna of Texas in the lower Rio Grande region. Feeling he had "exhausted most of the rare finds in that section," Armstrong traveled to eastern Mexico to collect in 1894, "with a set determination of bringing to science everything of interest." He collected bird eggs and made up taxidermic series of birds and small mammals. By 1902, however, he added "living wild animals and birds for scientific and propagating purposes" to the advertisement in which he described himself as a "collecting naturalist and dealer in Mexican and Southern Bird and mammal skins, bird eggs in sets." In 1905 the director of the National Zoo recommended Armstrong as a collector who "furnishes marsh and water birds of the southwestern United States and northern species which winter in that region. . . . Altogether [Armstrong] can get a large number of different kinds, and his prices are usually very reasonable." The zoo director did not entirely trust Armstrong, however, adding "it is best usually to have him guarantee safe delivery, and he is likely to be more prompt in filling orders if not paid in advance." Armstrong later turned to collecting reptiles, and adopted the nickname "Snake King."[28]

Other collectors in the natural history tradition also specialized in the fauna of the Southwest. J. W. Johnson, of Charco, Texas, advertised in 1913 as a "collecting naturalist and dealer in Mexican and Southern Bird and mammal skins, bird eggs in sets, living wild animals and birds for scientific and propagating purposes," specializing in armadillos, "fur bearers," and snakes. W. Odell in San Antonio dealt in "Texas and Mexican reptiles, parrots, and wild animals." These small businessmen and others supplied American zoos with North American animals.[29]

A traveling way of life, combined with experience in show business, offered another route to a career as an animal collector. Charles Mayer came from such a background. As a teenager, Mayer ran away from his boyhood home in Binghamton, New York, to join a circus around 1880. As he worked his way up in the circus world, Mayer traveled to

New Zealand, Siam, Singapore, Hong Kong, Japan, and Buenos Aires. In 1887 he decided to stay in Singapore and learn the animal-dealing business, apprenticing himself to a Malay dealer. He spent eighteen months in Sumatra living with "natives" and learning their language as well as their tracking and hunting techniques. For the next two decades he collected live animals regularly for the zoos in Melbourne, Perth, and Sydney, and occasionally shipped animals to Hagenbeck and the London dealer Cross. He also donated orangutan skins and an elephant skin to the Raffles museum in Singapore. Although he was American, Mayer had little contact with American zoos—import duties were too high, he explained, and the journey too long for safe delivery of the animals. Mayer wrote two books about his adventures, however, that were published and sold in the United States.[30]

When war broke out in Europe in 1914, German dealers dominated the import of exotic species to the United States, although some American animal dealers sold imported animals as well. These dealers made their living mainly from their pet stores, like Chester A. Lamb, of Dayton, Ohio, a "dealer in foreign and domestic birds," "monkeys, dogs, cats, and all pets, goldfish & etc." According to the National Zoo's director, Lamb was in 1905 a reliable supplier of "western quail, ducks, etc., and some of the more common mammals." On the West Coast, Ansel Robison's store, founded in San Francisco in 1852, was by 1900 an "importer and dealer in gold fish, birds, cages, animals . . . dogs, cats, monkeys." Robison's became a clearinghouse for animals arriving on ships from Asia. American fauna also was being collected, distributed around the country, and shipped abroad.[31]

The war disrupted this system. Germany lost its African colonies, and German shipping companies, which had been particularly accommodating to animal collectors' unusual cargo, ceased to operate. In addition, because of outbreaks of hoof and mouth disease in war-damaged countries, the U.S. Department of Agriculture's Bureau of Animal Industry placed restrictions on importing ruminants from all countries except England and Canada, in order to prevent the spread of disease to American livestock. This included all antelope, an important part of zoo collections. After 1919 ruminants could only be imported with a certificate signed by a veterinarian stating that the animal had been quarantined for sixty days before export. The Bureau of Animal Industry also set up a system of inspecting hoofed animals such as giraffes and zebras, as well as others, including elephants, rhinoceroses, and hippopotamuses. This bureaucracy curtailed casual gifts to zoos from overseas travelers. Animal dealing became to an even greater extent a specialty of businessmen rather than amateur enthusiasts.[32]

Despite constraints on imports, demand for animals increased. American zoos sought to replace animals that had died and to expand their collections. New zoos continued to open in the 1920s. In addition, other markets for exotic animals grew after the war. Wealthy wild-animal fanciers added to their private collections, such as the Hearst preserve at San Simeon, and the collection of Percy Godwin, of Nashville, Tennessee. Victor J. Evans, a patent lawyer who lived outside Washington, D.C., built up a private animal collection that included rare species of parrots as well as elands, bush pigs, zebra, and antelope. The pheasant fancier Colonel Anthony R. Kuser sponsored the collecting work and publishing of William Beebe, curator of birds at the Bronx Zoo. The number of private collectors of exotic animals is hard to judge. Individuals come to light, however, in letters like one from Joel W. Thorne of Millbrook, New York, to William Hornaday, in 1924. Thorne's efforts to breed zebra on his upstate New York farm had been unsuccessful, and he wrote to Hornaday for advice on how to sell the animals. In addition to private collectors, laboratories using rhesus monkeys provided a new market for animal dealers. Tens of thousands of rhesus macaques were imported for research on yellow fever and polio. The Hagenbeck firm reportedly sold eight thousand rhesus monkeys to the Rockefeller Institute in the 1920s for research on yellow fever.[33]

The setbacks to German animal dealers combined with new markets for wild animals allowed dealers based in the United States to expand, and new collectors and dealers entered the business. Some of the new dealers specialized in birds or reptiles. The warm climates of Florida and California, which made it possible to keep birds outdoors, made those states attractive to bird dealers. Florida, in particular, was a convenient point of importation for birds coming from Central and South America. A. E. Bascom, in Miami, for example, specialized in flamingos and Central and South American species. Bird Wonderland operated out of Van Nuys, California. Several reptile dealers also were based in Florida. Ross Allen, for example, started as a reptile collector, then opened an exhibit—his Reptile Institute in Silver Springs—which included an "Indian village, gift shops, museum, rattlesnake cannery, and biology supplies."[34]

John T. Benson was one animal dealer whose well-established, but modest, business took off after the war. Benson, born in England in 1872, traveled with the British Bostock Circus as a boy before coming to Boston in 1888. By 1895 he ran a private zoo at Norumbega Park, outside Boston, as well as a business exporting American birds. As the business expanded he moved to Lexington, Massachusetts, imported animals from European dealers, and exported a wider variety of American fauna. Around 1907 he moved again, to a 250-acre farm in Nashua, New Hampshire, returning to Boston for a brief stint as director of the Franklin Park Zoo in the late 1910s. Benson's animal-dealing business got a big boost in 1920 when he took over Sol Stephan's job as American agent for the Hagenbeck company. In 1922 Benson opened a "Wild Animal Show" at Coney Island. Apparently inspired by the success of this show, he developed his Nashua farm into Benson's Wild Animal Farm, where his animals for sale were on display. Crowds of up to forty thousand went to Benson's on Sundays to see the changing exhibits.[35]

Another animal dealer, I. S. "Trader" Horne, in Kansas City, Missouri, had opened his business a few years before World War I. Afterward, he became the main supplier of exotic animals in the Midwest. Henry Bartels in New York City, who had inherited a family-owned animal-dealing business dating from the 1890s, sent zoos extensive price lists of Indian and African mammals, birds, and reptiles in the 1920s. In the early 1930s Henry Trefflich, whose father had been manager of the private zoo of German animal dealer August Fockelmann, outside Hamburg, took over the Bartels business. An animal dealership run by Meems Brothers and Ward operated out of Oceanside, Long Island. All employed collectors in Africa, South Asia, and the East Indies. Although set back, the German dealers did not go out of business. Carl Hagenbeck's sons rebuilt the family enterprise, and Louis Ruhe maintained his New York offices.[36]

Along with other explorers of the natural world in the 1920s, animal collectors became celebrities. Books by Charles Mayer were published in 1922 and 1924, and J. L. Buck recounted how to catch a chimpanzee,

N'GI TAKES UP HOME IN ZOO

This 3-year-old gorilla, one of two in the United States, arrived at the Zoo this morning. He sobbed for an hour before he was released from his box to spring into the arms of J. L. Buck, his former owner. He posed for his picture with Mr. Buck.

GORILLA, SOBBING, UNLOADED AT ZOO

One of Two Now in U. S., First One Here, Embraces Captor When Freed.

A 3-year-old gorilla, the first ever at the National Zoological Park and one of the two now in the United States, arrived at the zoo this morning.

The astonishingly child-like behavior of the animal was impressed on the small group of zoo attendants and visitors as soon as he was released from the box in which he had been brought to Washington.

For an hour the little fellow had been sobbing audibly as if he had lost his last friend. For a year he has been brought up practically as a member of an American family, having the run of a large house, and confinement in a small box was an experience he could not understand.

Throws Arms Around Him.

Then the box was opened. The face of J. L. Buck, the animal dealer, who captured him more than a year ago in the twilight jungle of Spanish Guinea, on the West Coast of Africa, smiled down upon him. N'Gi sprang from the box, threw his long arms around Mr. Buck's neck, and began showering him with kisses as if he was a little boy. Buck stroked his black fur and in a moment he was pacified and willing to pose for his photograph.

Ordinarily a gorilla will not live in captivity more than a year or so. The longest record was 7 years in the Zoo at Breslau, Germany, but this never has been approached elsewhere. There the animal was brought up as a member of the zoo keeper's family and was treated as one of his own children.

It has been observed, however, that a baby gorilla usually will get along all right so long as he has the constant companionship of one human being to whom he becomes attached. Once this human leaves him for any length of time he becomes despondent and develops tuberculosis—that is, he practically dies of a broken heart.

Treats All Alike.

The animal now in Washington, while he has been reared as a member of Mr. Buck's family, has never been allowed to form a strong attachment for any one individual, but has been encouraged to treat all men and women as his friends. Hence it is hoped that he will transfer affections to the Zoo attendants and pass through the stage of inevitable despondency at the change without developing any organic trouble. If this period passes he may live for years.

N'Gi now weighs 35 pounds. When he reaches his full growth he will weigh between 400 and 500, but that will be a long time yet, since the gorilla attains his growth almost as slowly as a man. He is now in practically the stage of development of a 3-year-old boy. For a few weeks a guard will be kept at his cage night and day, never leaving him for a minute lest he wake in the night to find himself alone and injure himself in a paroxysm of grief and fright—possibly even commit suicide.

He has been placed in the big cage formerly the home of the giraffe with three other anthropoids—Jiggs, the orang-utan, and two gibbons. He has been raised with chimpanzees. Mr. Buck explained today, however, that the presence of the other anthropoids will not afford real companionship to N'Gi. He regards himself as a man and the others as monkeys. He will play with them, but only as a child might play with a monkey and not as his equals.

N'Gi is a gift to the Washington Zoo from Walter P. Chrysler, the automobile

(Continued on Page 19, Column 2.)

manufacturer, who financed the Smithsonian-Chrysler expedition three years ago.

The story of N'Gi's capture is a romance in itself. Starting from Lomie, an interior settlement in the British Cameron, Mr. Buck proceeded for five days through the jungles toward the Dja River. He was accompanied by a band of pigmies of the Batwe tribe, expert gorilla hunters and cannibals.

Finally the party ran upon signs that a gorilla family had passed that way. This could be told by the vegetation they had eaten. This trail was followed for five more days, when the family was located. It consisted of a male and four females, one of which carried the little fellow, then about 22 months old and still nursing at her breast, in her arms.

The hunters followed the family, pitching camp each night within a few hundred yards of their bivouac and following them as soon as they started out in the morning. The gorilla is not

Star - Dec 6

FIGURE 15. Animal collector J. L. Buck brought the baby gorilla N'Gi to the National Zoo. *Washington Star*, December 6, 1928. © *Washington Post*; reprinted by permission of the D.C. Public Library.

among other adventures, for *Asia* magazine. Two books were published about Sasha Siemel, who worked in South America. Siemel cultivated an image of Westerner gone native, hunting jaguar with a spear, although he made his living mainly by guiding hunting parties and shipping live jaguars to zoos.[37] Animal collecting stories made up part of a larger genre of tales of exploration and adventure that included the edited field notes of big-game hunters. They were also similar to the stories of naturalist-explorers like Roy Chapman Andrews, who traveled to Mongolia for the American Museum of Natural History, and Carl Akeley, the taxidermist whose work at that museum publicized the beauty and fragility of African fauna. The film makers Martin and Osa Johnson also brought public attention to the wildlife of Africa in the 1920s. Such explorer-celebrities occasionally donated animals to zoos. In 1931 the Johnsons, for example, gave the gorillas Ngagi and Mbongo, animals captured while they were shooting the film *Congorilla*, to the San Diego Zoo.[38]

The animal dealer who became most famous in the United States was Frank Buck. Like the careers of so many collectors and dealers, Frank Buck's livelihood stemmed from the combination of a rural boyhood spent catching snakes and trapping birds and a love of travel. According to his own account, Buck was born in a wagon yard near Gainesville, Texas, in the 1880s. At the age of eighteen he earned his transportation to Chicago working as a cowpuncher on a train load of cattle destined for the stockyards. In Chicago, Buck worked his way up from hotel bellhop to booker of vaudeville acts. He married a journalist named Amy Leslie and, by his account, spent late nights hobnobbing with theater folk in hotel restaurants. City life did not suit him, however, and, having won some money in a poker game some time around 1910, he decided to act on his wanderlust. Buck traveled first to South America. There he purchased birds and small animals, which he sold to the Bronx Zoo on his return to the United States. Buck made a second trip, to Brazil, and this time brought his cargo to London to sell. When the war in Europe shut down the German animal trade, Buck decided to move in on Hagenbeck's territory and left London for Singapore. He spent the rest of his animal-collecting career in South Asia and the East Indies.[39]

FIGURE 16. This advertisement for Frank Buck's display at the 1939–40 New York World's Fair promises "real jungle atmosphere." Author's collection.

Buck never lost the flair for show business he learned as a vaudeville agent. In 1915, back in San Francisco after delivering a shipment of animals, he worked as a publicity agent for the amusement zone at the Panama-Pacific Exposition. Later he parlayed his show business experience into promoting himself as an adventurer. Beginning in the 1920s he published eight books (all ghost written) about his "bring 'em back alive" exploits. He also made movies and hosted a radio show in the 1930s. He traveled with the Ringling Brothers and Barnum and Bailey Circus in the 1930s and made cameo appearances christening new buildings at the zoos in St. Paul and Baltimore. There were Frank Buck comic books, and Buck had his own exhibits at world's fairs—Frank Buck's Jungle Camp in Chicago in 1933–34 and his Jungleland at the 1939–40 New York fair.[40]

Collectors, Hunters, and the Ethics of Animal Adventures

Hundreds—perhaps thousands—of aspiring animal collectors responded to Frank Buck's publicity by writing to ask if they could join him on an expedition. Some of Buck's fans, however, seemed confused about what this would entail. One nineteen-year-old boy wrote, "I would like to go shooting wild animals with you. I would shoot enough to pay for my keep, being a crack shot."[41] If Buck's readers mistook him for a big-game hunter, it was perhaps because Buck looked like other male adventurers on the trail of wild animals wearing khakis and a pith helmet. Frank Buck's stories, however, were different from hunting stories in significant ways. The climax of hunting stories is the confrontation between the hunter and his quarry, and it is often told in bloody detail. Hunting is constructed as a metaphysical experience, a confrontation between man and nature that confirms white Western masculinity. Hunters, in theory, abide by a code of ethics with a notion of fair play: they shoot only the mature male animals that make worthy opponents. The techniques of animal collectors, by contrast, run directly counter to the sportsman's ethic. Animal collectors steal lion and tiger cubs while the female is off hunting, and they shoot nursing orangutan and rhinoceros mothers in order to capture their babies. Their goal is to capture baby animals, which are much easier to handle than adult males, and more likely to acclimatize to captivity.[42]

The decidedly unsportsmanlike conventions of animal catching, however, posed a problem for Buck, who used his narratives to raise his status as animal collector from marginal social figure to masculine adventurer-hero of popular culture. To solve this problem, Buck created his character not as a hunter, but rather as a heroic businessman. Buck rarely accompanied guides and trappers into the field to participate in collecting animals. He described this work—creating drives and pitfalls that snared the weak or young animals. But the work was done by his Malay assistants, stock characters referred to as "boys." Since catching animals alive was not a test of manhood like hunting, it was work suited to servants who were not acknowledged as men. Furthermore, whereas in hunting stories the human encounter with animals in the jungle formed the high point of the narrative, in Buck's stories capturing animals was just the first step in the process that allowed him to become a hero—keeping animals alive and transporting them to the United States.

Buck insisted that a collector's work was much more difficult than a hunter's, because capturing animals was more troublesome than shooting them, and once captive, animals had to be kept alive. It was no small task to make wild animals eat in captivity and to protect them from accidents and disease. The drama in Buck's stories takes place mainly during the time in between an animal's capture and its delivery to a zoo or circus. An astonishing percentage of animals died before reaching a collector's home port, so it was during this interval that Buck's investment in travel and in the upkeep of his animals was most at risk. Rather than a naturalist, hunter, or showman, Buck portrayed himself in his stories as a heroic businessman.

To be sure, the life of an animal dealer was dangerous, as Buck frequently reminded his readers. Buck's brushes with death usually occurred with wild animals in captivity. His job was more hazardous and more complicated than that of the hunter, Buck argued, because he preferred not to kill his opponent. In one story, as Buck was trying to maneuver a two-hundred-pound cassowary into a crate, the ostrichlike bird charged him:

> I backed away, tightening my grip on the stout bamboo pole. My problem was to smack that bird in the right place. If I hit him on the head or the neck (that stick of bamboo was about 2 1/2 inches in diameter and capable of dealing a terrible blow) I'd have a dead specimen on my hands. There is only one place to hit a cassowary if you want to stun him without ruining him—and that is on the side under the wing. . . . These things flashed through my mind as I poised myself to meet the bird's rush. As he sailed in I dealt him a healthy wallop—in the right place.[43]

Cunning, skill, and knowledge of his opponent's weak spot enabled Buck to both avoid personal injury and save his specimen.

The drama of Buck's books, however, lay primarily in his daring as an entrepreneur. Investing in wild animals was a big risk; they were likely to die before they could be sold, and they were costly to feed and house in the meantime. Buck lived the life of a businessman overseas, not an adventurer. He spent most of his time buying animals that had already been captured, dashing around to meet shipping schedules and drinking gin slings in the Raffles Hotel, the social hub for Westerners in Singapore. Once the cargo was assembled, Buck traveled with his animals, making sure they were fed and watered properly, and protecting them, in one story, from superstitious Chinese crewmen who wanted to kill his tigers for their gallbladders. When the crewmen succeeded in feeding the tiger rat poison, Buck's evaluation of the situation was businesslike: "I was out an animal that would have brought me not less than a thousand dollars."[44]

Buck could have described his work as the outcome of a desire to bring the wonders of nature before the American public, or as an effort to promote public concern for conservation. For example, popular books written for a German audience by Carl Hagenbeck's animal collectors attempted to portray their profession as rooted in a love of animals and nature. And by the 1920s museum naturalists and big-game hunters alike were well aware that wildlife populations in the East Indies and in East Africa had declined and that some species were near extinction. The British and other governments placed restrictions on the export of live animals and skins from their colonies, although they were not rigorously enforced. But Buck was concerned with such matters only when they presented a nuisance. He knew that the Indian rhinoceros was "practically extinct" in 1922, when he set out to procure two specimens of the animal, one each for the Bronx and Philadelphia Zoos. Yet he was unperturbed when the capture of his animals resulted in the killing of twenty-one additional rhinos, and even was amused by William Hornaday's horror at this turn of events.[45]

Frank Buck created his public persona, and his stories and movies became popular, at a time of increasing American political and economic power in the world. His writing brims with racial slurs and cultural stereotypes that readers a few decades later would find reprehensible. In the way he dressed, and in his treatment of local people during collecting expeditions, Buck modeled himself on the least sympathetic sort of European colonial official. By so doing, he explicitly associated the display of zoo animals with Western imperialism. In his 1985 novel World's Fair, E. L. Doctorow deftly captured Buck's unpleasant personality as well as the persistent public fascination with the kind of power he represented. After visiting Frank Buck's Jungleland at the 1939–40 World's Fair in New York, the novel's protagonist reflects:

> The truth was, I thought now, Frank Buck was a generally grumpy fellow, always cursing out his 'boys' or jealously guarding his 'specimens' or boasting how many he had sold where and

for how much. He acted superior to the people who worked for him. He didn't get along with the authorities in the game preserves, nor with the ships' captains who took him on their freighters with his crated live cargo, nor with the animals themselves. I saw all that now, but I still wanted to be like him, and walk around with a pith helmet and a khaki shirt and a whip for keeping the poor devils in line.[46]

For different reasons American zoo directors also had mixed feelings about Buck. They were grateful for the publicity Buck brought to zoos, but called him a "faker" because he was a businessman rather than an animal man. Zoo men pointed out Buck's inexperience in animal keeping by circulating a story about how Buck was fired from a stint at the San Diego Zoo, because he made the elephants ill by treating their skin with oil. Knowledge of animal keeping was prized in the zoo profession, and animal collectors other than Frank Buck earned respect for their expertise.[47]

Collectors and the Craft of Animal Care

An animal collector's livelihood and reputation rode on his ability to transport animals in safety and good health. Some collectors gathered their animals at holding stations before transporting them so that they became accustomed to captivity, a new diet, and handling by humans before being subjected to the shock of life in a cage lashed to the deck of a ship. If a collector could not travel with his animals to oversee their care, he depended on the ship's crew to do this work or hired someone for the job. An animal dealer based in Buenos Aires, Argentina, for example, regularly shipped his animals to the United States aboard the S.S. *Brazil* because he could rely on the smoking room steward, Alfred Bienenwald, to take good care of his cargo.[48]

Even in good circumstances, the loss of life was astonishing. No firm figures exist on the percentage of mammals, birds, or reptiles to survive the journey from jungle to zoo, but anecdotal evidence is suggestive. Hagenbeck reported that half of the animals in Cassanova's "Nubian" caravan in 1870 had succumbed during the fifty-four-day journey from the interior of Africa to the Red Sea port of Suakin (from there the animals had been transported by steamship to Suez where

Hagenbeck met them). Frank Buck recounted a journey he made, accompanying a Calcutta bird dealer, to a village in the foothills of the Himalayas. Tribal people there captured pheasants rarely seen in the West—blood pheasants, tragopans, Impeyans, and others—and brought them to the village to sell. The bird dealer bought two thousand birds, of which Buck estimated that ten percent were already dead. "Most of them were packed so tightly in the baskets and crates that it was impossible for them to move," Buck explained. "Some of them were merely wrapped in rattan, the basket actually woven around the bird so that it could move neither legs nor wings . . . many of them had been in these baskets and crates for from ten days to two weeks." By the time Buck and the bird dealer got back to Calcutta half of the remaining birds had died. Of the three hundred he chose from this lot to ship to the United States, 125 arrived in San Francisco alive. If the bird dealer's stock shared a similar fate, perhaps fifteen percent of the original two thousand birds survived approximately four months in captivity.[49]

Keeping the precious animals that arrived at the zoo alive in captivity was an urgent problem at the turn of the twentieth century. Animal keeping was a craft; the knowledge of how to care for wild animals in captivity was not codified. Diets and animal management practices were not usually based on knowledge of the animal's life in the wild. Even in cases where it was known what an animal ate in the wild, this diet was not necessarily duplicated in captivity.

Animal collectors were part of the network of people who shared and contributed to the knowledge of keeping wild animals in captivity. Carl Hagenbeck was well known for his efforts in acclimatizing tropical animals to the cold damp winters of northern Europe. Whereas nineteenth-century European zoos kept lions, tigers, giraffes, ostriches, and monkeys in heated houses, Hagenbeck put these animals and others outdoors during the winter, in paddocks where they could exercise, and with access to shelter if they wanted it. Hagenbeck described his experiences in his memoir, concluding that, "No fallacy is more widespread than that wild animals have to be kept throughout the winter carefully guarded from the effects of the low temperature."[50]

Zoo directors also had to be alert to misinformation on animal care. The director of the National Zoo referred the director of the Philadelphia Zoo to an article that, "I think you will find . . . rather amusing as it is an article on the care of animals and the construction of quarters for the same by a man who has never had anything to do with either." The first comprehensive, and widely circulated, book on the care of wild animals in captivity was published in 1950.[51]

Zoos and the Culture of Collecting

American zoos depended on a well-established network of animal collectors and dealers to supply their animals. Animal collecting was a business—a form of colonial commerce dependent on the social structures and lines of transportation set up by European powers in Africa and Asia. A variety of cultures, however, had a stake in bringing live animals from their wild habitats to more domesticated settings. The demand of circuses for certain animals influenced what species collectors focused on. The circus world inspired some young men to set out on an animal collecting career. Others started collecting from an interest in natural history. Either way, collectors shared a common knowledge of how to keep wild animals in captivity. This knowledge enabled them to go back and forth between the worlds of the collecting business, circus and vaudeville entertainment, and zoos throughout their careers. The constraints under which collectors worked—the accessibility of places where wild animals could be caught, the demands of the market, individual skill at managing both their animals and their business—influenced the availability of animals to zoos. The natural history of the animals themselves also played a role in this process: animals that did not acclimatize to life in captivity died.

Without animal dealers, zoos would have taken a different form, if they existed at all. Displays would have been limited to local fauna, deer, and birds. But the trade in wildlife gave zoos access to a wide variety of animals. By 1895 the Cincinnati Zoo advertised exhibits typical of a large and successful collection:

> The llama, deer and elk parks; the great monkey castle, with its hundreds of chattering and grimacing Simian specimens; the lake, with its many curious birds and water fowl; the eagle houses; the owl cages; the zebu, yak and buffalo paddocks; the

giraffe and elephant houses; the villages of the prairie dogs and ground hogs; the sea-lion basin; the wolf and badger dens and fox houses; the bear pits, with their huge polars and grizzlies; the camel yards; the emu and ostrich enclosures; the carnivora building, with its lions, tigers, leopards, wild cats, hyenas, etc.; ... the aviaries, with the hundreds of vari-colored birds of the tropics; and the thousands of other bird and animal attractions, together with the pony and elephant and camel tracks.[52]

Through the commercial animal trade American cities became populated with elephants, polar bears, and flamingos, and zoos became one among many institutions taken for granted as part of urban life.

CHAPTER 4
ZOO EXPEDITIONS

When William M. Mann became director of the National Zoological Park in 1925, the zoo had never exhibited a giraffe, and it owned no rhinoceros. It had received a few gifts of rarities but lacked many of the large mammals that were considered essential to a zoo. To boost the zoo's collection, Mann began lobbying for an animal collecting expedition. Within a year, he was leading a collecting party to British-ruled Tanganyika in East Africa, financed by the car manufacturer Walter P. Chrysler. With a budget of fifty thousand dollars, it was the first large-scale expedition mounted by any American zoo, and it was the first of three major collecting efforts led by Mann.[1]

Mann's goals during his zoo expeditions differed from his goals on expeditions earlier in his career, when he collected ants as a museum entomologist. As an animal collector, Mann did not seek to advance his own professional status through the mechanisms of the scientific community, by publishing technical papers or using the collection for further study once it was home. In addition, although Mann's general goal in zoo collecting was to "add to our knowledge of life," the audience for that knowledge was not primarily the scientific community. Rather, the intended products of Mann's expeditions were popular: newspaper and magazine articles and photographs, lectures, films, and, most important, the display of the collection in a public park.[2]

Publicity surrounding Mann's collecting expeditions gave zoo visitors incentive to go to the park and encouraged them to imagine a trip to the zoo as an expedition. A zoo visit was billed as an escape to nature, and one aspect of looking at the animals was for zoogoers to travel in their minds to the exotic landscapes and cultures from which the animals had journeyed. Newspaper articles and radio broadcasts about Mann's field work gave visitors the vicarious experience of zoo expeditions—of participating in the work of an elite scientific institution—and provided a context for creating the illusion of an expedition at the zoo. Some zoos made the connection more explicit, organizing zoo tours as "safaris." By enhancing the experience of a zoo visit, the stories that Mann collected on his expeditions were just as important as the animals.[3]

Mann's expeditions resembled those of the National Geographic Society, which also were justified by public interest rather than the research program of professionalized science. In fact, the National Geographic Society funded one of Mann's expeditions, and Mann contributed several articles to its magazine in the 1930s. Popularizations were not merely by-products of National Geographic expeditions—rather, the trips were planned with them in mind. In reporting expedition results, tales of adventure were valued as much as natural history description. The first-hand reports of National Geographic explorers gave the society's members a sense of participating in science (even though exploration in itself held little status among professional geographers as a scientific endeavor by the early twentieth century). These expeditions set out to do popular science, rather than to popularize elite science.[4]

Likewise, vicarious public participation was built into Mann's expeditions. The 1926 expedition to Tanganyika, for example, was covered by a newsreel photographer from the Pathe Review; during Mann's absence, the *Washington Star* sponsored a contest for children to name the giraffe that Mann promised to bring home; a representative from the Smithsonian read Mann's letters from abroad over the radio; and in the months after the expedition Mann gave dozens of lectures about it. Mann's expeditions, furthermore, were carried out with the privileges of affiliation with a scientific institution—the Smithsonian Institution. As a reputable zoologist leading a scientific expedition, Mann was treated as a dignitary in foreign countries. Scientific status helped elicit cooperation from U.S. State Department personnel and smoothed access to colonial officials in the countries Mann visited. In addition, as a scientist Mann obtained export permits that were not available to commercial collectors or to hunters.

As enterprises with scientific status and popular goals, expeditions to collect animals for the National Zoo provide an opportunity to study the meaning of popular science as it is constructed in the field. These expeditions shed light on the practices that brought natural objects into the popular forum of the zoo. Putting animals on display at the zoo was the culmination of field work. At a minimum it required capturing, purchasing, caging, trading, and transporting wild animals. Like the work of other scientific expeditions, collecting animals for zoos was shaped by local conditions: conventions of travel, the development of tourism, and the services and social structures of colonial culture. In addition, animal collecting relied on networks of local collecting expertise and of commercial animal dealers. News of the National Zoo's expeditions that reached the public focused on the adventure of field work—both the work of animal collecting and the social relationships between Mann and others in the field.[5]

William M. Mann's major expeditions to collect animals for the National Zoo—to Tanganyika in 1926, to the Dutch East Indies in 1937, and to Liberia in 1940—differed in organization, and local field conditions had a significant impact on Mann's success in collecting animals. Mann's skills in securing trustworthy help, locating and capturing animals, and feeding and transporting them proved effective in Tanganyika and the Dutch East Indies, where an existing tourist industry, regular steamship service, and local animal-catching expertise facilitated the work. Collecting was more difficult in Liberia, which lacked an infrastructure to support collecting by Western naturalists. Popular accounts of Mann's expeditions focused on the work of animal collecting and negotiating through foreign cultures in order to collect. The animals Mann brought to the zoo were souvenirs of this larger enterprise—expedition work—which was also brought before the public. Even when Mann was not successful in obtaining the animals he wanted, the stories he brought home made the trips worthwhile.

Why Mount an Expedition?

It is important to understand why Mann undertook the job of collecting himself. William M. Mann was nearly forty years old when he became director of the National Zoo. Trained as an entomologist, he had spent his entire career collecting. As an undergraduate at Stanford University he accompanied entomological expeditions to Mexico and Brazil. During doctoral studies at Harvard, under the renowned ant authority William Morton Wheeler, Mann collected ants and termites in Haiti, Mexico, Palestine, Syria, Fiji, the British Solomon Islands, and Australia. Between receiving his Ph.D. in 1916 and being hired by the zoo, Mann worked as an entomological explorer for the U.S. Department of Agriculture and traveled in Central and South America.

Mann bore a physical resemblance to Rudyard Kipling, and friends recalled him as a colorful personality, kindly and generous, "small in stature, wiry, careless of dress, with a puckish expression, and a keen sense of humor." In 1922 Mann had accompanied an expedition to the Amazon basin sponsored by a chemical company; a popular book about the trip immortalized him as wearing "the most disreputable hat in North and South America," and was dedicated to him, "the bug-hunter, stout companion of the trail." Mann also had loved zoos from boyhood, and he returned from the Amazon expedition with more than one hundred live animals, which he gave to the National Zoo in Washington, D.C.[6]

When Mann became zoo director, however, he looked at the institution he inherited with the eyes of a collector who had amassed, classified, and mounted tens of thousands of insect specimens (his personal collection numbered 117,000 when he donated it to the Smithsonian in 1955). What Mann saw was far from his ideal of "a representative collection of live wild animals from all parts of the world."[7] Yet Mann had more in mind than filling out the zoo's animal collection like a museum series. Given the extensive commercial animal trade, Mann could have purchased many—although not all—of the animals he wanted from dealers and saved himself the trouble and expense of expedition work. When the Dallas Zoo started up, for example, the dealer Frank Buck supplied all the animals. But Mann had reasons both personal and practical for going into the field. First of all, he loved to travel; he was an experienced field naturalist, and during his entomological expeditions Mann had picked up live animals that he donated to zoos. Furthermore, Mann had negotiated the possibility of leading collecting expeditions before he took the job as zoo director.[8]

There were also reasons to be wary of buying animals from dealers. As with rare objects in the antiquities business, the source and condition of an animal described on a dealer's price list was always open to question. Putting money down on an animal before seeing it was risky. As Gustave Loisel explained in his encyclopedic 1912 *Histoire des Menageries*, buying from dealers was convenient, but—particularly in the case of rare animals—the pressure to make a deal quickly put the buyer at a disadvantage. In any case, animals often arrived in poor health. William Temple Hornaday, director of the New York Zoological Society's Bronx Zoo, warned a friend to be wary of dealers: "There are plenty of fly-by-night animal dealers in the country, who deal in animals only on paper, such as Garland, of Oldtown, Maine, and others I could name."[9]

Furthermore, reputable dealers did not supply animals that were fragile or did not survive well in captivity. For a commercial dealer it did not pay to collect animals that might not survive to be sold, such as pangolins (spiny anteaters) or mouse deer. A zoo-sponsored expedition could take that risk, and also could invest more in the care of animals on their months-long ocean journey to America. In addition to coddling such rarities, a zoo-sponsored expedition could save money by eliminating the middleman of the animal dealer.[10]

Even more important than fun and an opportunity for one-upmanship with respect to other zoos, Mann saw expedition work as part of institution building, facilitating his goal to "establish and maintain wide contacts" for the zoo. By acting as an ambassador between the zoo and people abroad who were concerned with wildlife—local collectors and dealers, game managers, diplomats concerned with hunting or conservation, zoologists, and zoo directors—Mann sought to build a network of contacts analogous to those of the Smithsonian Institution's National Museum. Collecting and networking were activities in keeping with other branches of the Smithsonian, and Mann's status as a representative of both the Smithsonian and the federal government helped him in these tasks.[11]

Finally, and above all else, expeditions made effective public relations vehicles. They generated public enthusiasm for the zoo (as evidenced by increased attendance when expedition-caught animals were put on display) and brought its patrons to public attention. The amount of publicity an expedition could generate initially caught Mann off guard. In particular he was surprised when the Pathe Review, a newsreel company, jumped at the chance to accompany him; Mann considered "African moving pictures" already in 1926 to be "rather a drug on the market."[12] Although it had been part of the Smithsonian for more than thirty years, the National Zoo in the 1920s did not have a national reputation, and was often referred to as the Washington Zoo. Mann's 1926 expedition, through the combined public relations efforts of the Smithsonian and Walter P. Chrysler, changed this. William Mann described the effect of the announcement of the expedition enthusiastically: "The advertising value to the Zoo is simply enormous. We may be certain that in today's editions of more than one thousand American newspapers there is an account of what we intend to do, and I would wager that some thousands of people who never knew that there was a National Zoo in Washington, learned about it this morning."[13] National awareness of the National Zoo was important because the zoo depended on Congressional appropriations for its budget. Expeditions thus not only supplied animals, but also brought a variety of institutional benefits to the zoo.

The Smithsonian–Chrysler Expedition to Tanganyika, 1926

In April 1926, the British ship S.S. *Llanstephan* voyaged eastward across the Mediterranean, then through the Suez Canal and the Red Sea, on its way to the east coast of Africa. By coincidence, the ship carried the personnel and provisions of four different expeditions with an interest in African wildlife. In addition to the Smithsonian–Chrysler Expedition, there were two groups of sportsmen with plans to hunt big game in Kenya ("they have an elaborate armory and enough ammunition to equip a small company of soldiers," observed a fellow passenger), as well as the Akeley–Eastman–Pomeroy African Hall Expedition for the American Museum of Natural History, perhaps the most lavishly outfitted expedition sent by any museum to Africa. The three expeditions collecting dead specimens disembarked in Mombasa to test their skill in Kenya Colony. Mann and his party continued south to Dar es Salaam, in Tanganyika Territory (now Tanzania), also under British rule.[14]

Africa was a logical choice as a place for Mann to lead his first zoo expedition. Few African animals had been imported into the United States since before World War I, and the National Zoo's collection of

African animals was depleted and aging. Just as important, the fauna of Africa represented wildlife in general to Americans; Americans considered Africa the "world's zoo." An earlier Smithsonian expedition, led by Theodore Roosevelt to East Africa in 1909, had contributed to that impression. The so-called Smithsonian African Expedition sent more than ten thousand specimens of mammals, birds, reptiles, and fishes, and several thousand plants to the Smithsonian's National Museum, as well as several dozen live animals to the National Zoo. Roosevelt's books about the expedition remained widely distributed through the 1920s and 1930s. The animals that Mann most urgently desired—giraffe, rhinoceros, and elephant—were icons of the African wilderness. Mann decided to go to Tanganyika for these animals because he considered the Sudan, South Africa, and Kenya "collected over," and Abyssinia too far from the coast for transporting live animals.[15]

According to the secretary of the Smithsonian, Mann's expedition was to be "the most extensive expedition to be undertaken by the Smithsonian since the Smithsonian–Roosevelt Expedition." The romance of Roosevelt's big-game hunting fired Mann's enthusiasm. "All of us," he wrote, "in our earlier imagination travel in Africa, slaughter elephants, dodge spears, and suffer delightful hardships." Mann was not the only one excited by the prospect of the trip. Dozens of boys and men from around the country wrote to Mann asking if they could accompany him, after the expedition was announced through a press release to the nation's newspapers.[16]

Mann's expedition to Tanganyika was a combination of private hunting safari, government-sponsored scientific expedition, and publicity stunt in its personnel, equipment, and organization. Many wealthy sportsmen in the 1920s offered their patronage to natural history museums and accompanied the expeditions they sponsored. In exchange they could claim to be on scientific expeditions and hunt with exemption from colonial game laws; the museum would receive a few animal skins. Rather than his patron Chrysler, however, Mann was accompanied in the field by two close friends who had no affiliation with the zoo: Frederick Carnochan, Mann's friend from graduate school and a specialist in beetles who was to help with collecting, and Stephen Haweis, an artist and naturalist whose task was to care for the animals once they were captured. The newsreel photographer ensured that the expedition would receive publicity.[17]

As far as the British colonial government was concerned, the zoo expedition was in fact subject to the same rules as private hunting parties until proven a scientific expedition. Under pressure from conservationists, game laws by the 1920s required hunters and collectors to purchase expensive licenses and limited the numbers of animals that could be killed. A fourth expedition member, Arthur Loveridge, was particularly helpful in clarifying to the British government that Mann was leading a collecting expedition for science rather than a recreational safari. Loveridge, a reptile specialist who worked at Harvard's Museum of Comparative Zoology, had served as an assistant warden in the Tanganyika Game Department for several years. With his knowledge of game laws and his recent experience in Africa, he was adept at drafting letters, to be submitted through the U.S. State Department, requesting special permits to collect. The letters emphasized that "It will be a scientific expedition, and game animals will not be hunted any more than is necessary to obtain living young." As a scientific expedition, the zoo expedition was exempt from paying the fees for licenses and avoided import duties on its equipment.[18]

Loveridge also traveled to London ahead of the rest of the group to make additional purchases required of expeditions to British colonies: hats, shirts, and trousers for the 150 African porters Mann expected to hire. In addition, the expedition carried crates for the animals they expected to capture, and chairs, cots, nets, pistols, trunks, mess kits, and tool boxes supplied by the U.S. Marine Corps, as well as a Chrysler truck, specially outfitted for the trip in the company's London factory.[19]

Once in Tanganyika, Mann set up a base camp 250 miles inland at Dodoma. Loveridge stayed at the camp while the others split up. Carnochan and Haweis evidently had success in trapping animals and organizing drives to catch them. Within three months they had assembled fifteen hundred specimens, including "hyena, giant civit cat, a magnificent pangolin, a rare wild cat, forty monkeys, a dozen antelope and a really fine collection of birds."[20] In the meantime, Mann struck out for big game by the same methods as other game enthusiasts who arrived in East Africa knowing few people and speaking no Swahili.

Like sportsmen on safari package tours, he hired a "white hunter" as a guide and to manage the porters. Mann's hired companion, Charles B. Goss, was to help him capture rhinoceros and giraffe. The rhinoceros proved elusive. Mann's letters home detailed the hardships of camp life that made hunting stories so appealing: "We saw tracks every day ... I shall remember Africa as a place full of rhino trails to follow. We camped near to and far from drinking places and waited both early in the morning and late in the evening. We built platforms in trees above the water holes and watched all night. Any amount of game came to drink, impalla, wildebeeste, congoni, and so forth. A lion circled us half a dozen times and three hyaenas came directly beneath us. But there were no rhinos."[21]

Mann was not about to leave Africa without a giraffe, however. The giraffe had been touted as the main object of the expedition, and newspapers had celebrated Walter Chrysler as a friend of children for donating the funds to secure one. The *Washington Star* eagerly awaited news that one had been caught in order to begin the naming contest. To pursue giraffe, Mann and Goss engaged the assistance of two tribal leaders and four hundred of their men. For days they attempted to drive giraffe herds into 200-foot-long nets, but "in each case the big bull that headed the herd dashed through the nets, the others following in single file." They also tried building a corral, with little success, but eventually a young giraffe that became separated from its mother was surrounded and captured.[22]

The giraffe died before it could be transported home, however, and in the end Mann purchased two giraffes from the government of Sudan on his way home. It did not matter to Mann's patron or his audience that he had not captured the animals himself. He had already transmitted home stories of life in the field in pursuit of giraffe. These stories, which were broadcast over the radio, were just as valuable a product of the expedition as the animals themselves. Although Mann returned without the purported object of the expedition—a wild-caught giraffe—the stories of pursuing giraffes served similar publicity purposes.

The expedition returned to the United States on October 24, 1926 with more than twelve hundred animals. It was a huge public relations success. The media praised the expedition's sponsor, Walter P.

Chrysler. And the giraffes purchased in Sudan, a male and a female named Hi-Boy and Dot, attracted record-breaking attendance at the zoo. The Smithsonian Institution, which happened to be in the middle of a campaign to raise endowment money, also received favorable publicity. In addition, Mann achieved national recognition as a naturalist. Soon after the expedition, the Science Service produced a series of records for classroom use featuring seven prominent scientists lecturing on their areas of expertise. Mann, whose topic was "Our Animal Friends," was given equal billing with Nobel Prize winner Robert Millikan on "The Rise of Physics."[23]

The institution of the big-game safari had shaped the expedition from start to finish. Mann chose collecting grounds on the basis of the reputation of African game made popular by hunters. The rules and regulations for hunting in Tanganyika and the practice of big-game safaris laid out a procedure by which the expedition could locate and capture animals. Two people in particular played important roles in facilitating the expedition—Arthur Loveridge, with his game warden expertise, and Charles Goss, the "white hunter." Such people, who could guide expedition members in a foreign country and culture, proved even more important during Mann's expedition to the Dutch East Indies.

The National Geographic Society–Smithsonian Institution East Indies Expedition, 1937

On January 19, 1937, William Mann set off on the National Geographic Society–Smithsonian Institution East Indies Expedition to collect animals for the National Zoo. Funds from the Public Works Administration had recently made possible a new house for small mammals and great apes at the zoo, as well as an elephant house, and an addition to the bird house. Although not complete, the work was well under way—cage space had been allotted, pedestrian paths designed, foundations dug, and concrete poured. The new houses needed inhabitants. Mann was relieved to escape the "struggling with architects and artists," as he put it, and get back to his real love: collecting specimens in the field.[24]

Mann was accompanied by his wife, Lucile Quarry Mann (they had married shortly after Mann's return from Tanganyika), an experienced journalist who kept a detailed diary of the trip. Maynard Owen Williams, a photographer and chief of the foreign editorial staff of *National Geographic* magazine, traveled with the Manns. The zoo's highest-ranking keepers, Roy Jennier and Malcolm Davis, voyaged separately, joining the expedition in Sumatra with a cargo of American mountain lions, alligators, opossums, raccoons, black bears, jaguars, and hellbenders (large salamanders) to be dispensed as good-will gifts.[25]

Mann first proposed the expedition to Gilbert Grosvenor, president of the National Geographic Society, in 1934. Grosvenor responded with enthusiasm; the National Geographic Society contributed $17,000 toward the expedition, "as part of its work for the increase and diffusion of geographic knowledge." Again, an important product of the expedition was coverage in the popular press. As stipulated by Mann's contract with the Society, Mann agreed to write articles for *National Geographic* magazine and to keep in touch with the Society's Washington, D.C., headquarters during the expedition so that it could prepare and distribute press releases describing his progress.[26]

The East Indies had both natural and cultural resources that facilitated the work of animal collecting. Including, for the most part, the islands that are now Indonesia, the territory stretched from Sumatra to New Guinea, a distance of nearly four thousand miles (fourteen days by steamer). Long before the Portuguese and Dutch entered the spice trade in the sixteenth and seventeenth centuries, these islands had been stopping points on the trade route between India and China. Natural products of the East Indies, including rhinoceros horns and other materia medica, edible bird nests, and shark fins, formed part of this trade.[27]

In the nineteenth century, the East Indies became increasingly popular collecting grounds for Western naturalists. Sir Stamford Raffles shipped samples of the local flora and fauna back to England early in the century. Dutch and German naturalists collected specimens and studied them, for example, at Buitenzorg, the Dutch research station on Java. Alfred Russel Wallace made his famous journey through the Malay archipelago in the 1850s. William T. Hornaday, who later became director of the Bronx Zoo, was in the 1870s a taxidermist collecting in the East Indies for Henry Ward's Natural Science Establishment. Collectors of live animals went there too—they particularly coveted orangutans and other primates, tapirs, boa constrictors, and birds of paradise.[28]

The logistical hurdles of obtaining collecting permits aside, the well-known fauna of the East Indies and the ease of travel to and around the islands made them a good destination for a zoo expedition. While planning the expedition, Mann was happy to find that "The Dollar Steamship Line has a $745 round-the-world trip," and that on Sumatra, "There are over 700 kilometers of railroad . . . a number of bus lines, and the place is said to be full of Malay and Chinese auto drivers, so there is plenty of local transportation." From "a Zoo standpoint," Mann wrote, "Sumatra is as good a place as one could go for a short stay." Other naturalists had the same idea; Sumatra in 1937 became a crossroads of Harvard-affiliated expeditions. The primatologist Harold J. Coolidge, Jr., who had been studying gibbons in Siam, met the Manns there, as well as two other members of his expedition from the Museum of Comparative Zoology, Barbara Lawrence and Clarence Ray Carpenter. Harvard entomology professor Charles Thomas Brues and his wife Beirne Brues, a botanist, also joined the Manns for part of their expedition. Mann, in fact, had tried to persuade his mentor William Morton Wheeler to join the expedition, but Wheeler declined.[29]

By the 1930s, the East Indies were so well covered by collectors that the Dutch colonial government had taken measures to protect the area's wildlife. In 1916 the governor-general of the Dutch East Indies was given the power to designate territory with particular scientific or aesthetic significance as nature reserves. By 1930 there were seventy-six such reserves, mainly in Java, ranging in size from more than six hundred square miles to a single protected banyan tree. In general, in the 1920s and 1930s, colonial governments began regulating the hunting and collecting of wildlife, recognizing that many species were being pursued to extinction.[30]

To prepare for the expedition, Mann took advantage of diplomatic channels as well as other social networks created by Western imperialism in the East Indies. He wrote letters to Dutch and British colonial officers and U.S. State Department officials in the East Indies, as well as Borneo, the Malay peninsula, and India, to obtain permits for

collecting. From the Dutch East Indies Mann requested permits to collect one breeding pair each of various protected species, including
orangutan, two species of gibbon, rhinoceros, tapir, babirussa, and
anoa (dwarf buffalo), as well as hornbills, birds of paradise, and casso-
waries. In addition, Mann wrote letters to people he knew in Bangkok
about hiring local collectors, and the possibility of getting a rhinoceros
in India. He approached the Goodyear Rubber Company for assistance
from its East Indies plantation. The U.S. Secretary of State and the Secretary of the Smithsonian Institution prepared letters of introduction
for expedition members.[31]

Mann also got in touch with friends who had collected recently in
that part of the world. Harry Wegeforth, president of the Zoological
Society of San Diego, warned him that commercial animal dealers were
creating problems for scientific expeditions. Wegeforth wrote, "Frank
Buck also has been a factor in making it more difficult to get animals
out, because there is an impression among the people of the East that
he has made millions of dollars on the animals he has brought back to
the United States." Tapirs, he said, were impossible to get out of the
country, and Mann should beware of new import-export duties being
charged in Singapore.[32]

William and Lucile Mann, accompanied by the *National Geographic*
photographer Maynard Williams, boarded their ship for the East Indies in Seattle. They did not journey directly to Sumatra, however.
They first carried out some of the networking that Mann considered
so important to the scientific mission of the zoo. They stopped in
Tokyo, then Kyoto and Osaka, Shanghai, Hong Kong, and Singapore
to tour the local zoos and meet with the directors, local naturalists, and
animal dealers. On March 2, 1937, the Manns and Williams arrived in
Belawan on the north coast of Sumatra. They immediately drove fifteen miles to Medan, the regional capital, to call on the American Consul regarding problems with their collecting permits, which dogged
the expedition into mid-May. On a trip to Batavia (now Jakarta, the
capital) in April to negotiate with the American Consul General, the
Manns discovered for themselves just how crowded with commercial
collectors the Dutch East Indies had become. Lucile Mann named some
of them in her journal: "We were astonished to find Batavia had almost
as many animal collectors in town as animals. On the boat in the morn-

ing we had met Danesch, collecting for Amazonica in New York. In the Zoo we learned that Meems, of Ward and Meems, and Kreth, of [Louis] Ruhe, were here, too."[33]

While waiting for their permits to obtain protected species, the Manns set about assembling a collection of animals that were not protected. Collecting animals involved much more than setting out traps in the jungle; it was a social activity requiring participation from the highest government officials down to locally hired servants. It meant making contact with Dutch naturalists who could help them, recruiting native people to collect animals for them, and setting up a base camp. One of the first people the Manns got in touch with was a Dutch veterinarian named J. A. Coenraad. Coenraad was also a collector and ran a small zoo. He and his wife proved to be not only pleasant company and excellent tour guides, but also important social contacts. Coenraad introduced the Manns to local naturalists and rubber company officials, and he helped the Manns get started collecting.

Most animal collectors did not learn of animals' whereabouts and habits by personal observation. The most efficient way to collect animals—and one that was well established—was to take advantage of the knowledge and experience of native people. William T. Hornaday gave this advice to an aspiring collector in 1924: "Most collectors obtain monkeys, parrots, and macaws from the natives, who know how to capture them. If, when you locate in your chosen jungle, you send out word far and near that you will buy live specimens at good cash prices, and then establish a reputation as a good sport by occasionally buying something that you really do not want in order to encourage the trade, you will soon have the jungle people coming to you from near and far."[34] As soon as the Manns arrived, Coenraad had sent out word to his native collectors. In a few days, two specimens of the smallest wild cat—*Felis minutae*—and two parrots had been brought in.[35]

The Manns set up a base camp in an abandoned hospital on a rubber plantation. The location was isolated enough to qualify as a quarantine station; to guard against hoof and mouth disease, ruminants and swine would have to be held sixty days before they could be shipped to the United States. William Mann described camp life: "We have a Zoo organized comparable to that in Washington—something as follows: Lucile Mann, secretary, and Layang Gaddi of Borneo, purchasing agent of provisions; animal force, Jennier and Davis, and two harum scarum boys as assistants; carpenter department, Wing Hap and assistant; police department, Ram Singh, night watchman."[36] The Manns also hired a housekeeper and a "boy" whose sole duty was to catch grasshoppers for bird food. Rental of the twelve-acre property also included salaries for five gardeners and a night watchman. They had electricity installed in the building, rented an icebox, and established an account at the local Chinese grocery store.[37]

The morning after the Manns' first night at their camp, local people began coming by with animals. A Chinese man from the northern Sumatran province of Atjeh became a regular supplier. On his first visit he brought a wild dog, a martin, an otter, a hog badger, a loris, and various birds. Another man brought a baby tiger that had to be fed milk from a bottle, and became one of the camp's favorite pets. One collector was nicknamed the "hornbill specialist" for contributing so many of these birds. Davis and Jennier, the zoo keepers, set out some traps but had little success with them. They did capture the occasional baby civit cat or snake that wandered into the compound. Local markets provided another source of live animals. In some towns specialized bird markets supplied pets.[38]

Culture Brokers

The work of animal collecting in the East Indies depended on the help of cultural translators who could smooth the way for the Manns: people who spoke English as well as the Malay language and local dialects, people who knew where and how to hire local carpenters and other camp help, people to whom the Manns could delegate responsibility for day-to-day tasks that were difficult in a foreign culture. In the literature on colonialism such people are known as culture brokers. They might be young native people educated in missions who help missionaries become established in local culture. In Africa, locally hired assistants were essential to the field work of Western anthropologists. A culture broker need not be a native person. J. A. Coenraad, for example, helped the Manns pursue their interests by guiding them through the customs and policies of the Dutch colonial government.[39]

The most important person the Manns hired locally was a man described as a "Borneo Dyak," named Liang Gaddi Sang. (The Dyak are

FIGURE 17. William Mann bargains with children in Indonesia to purchase their pet. Maynard Williams/National Geographic Society Image Collection.

FIGURE 18. Expedition members bought birds at a market in Soerabaia. Maynard Williams/National Geographic Society Image Collection.

the indigenous people of Borneo, reputed to be head hunters, a notoriety exaggerated in the writing of early Western travelers to the region.) Gaddi, as he was called, had a well-established reputation as a collector and taxidermist. Born in Borneo, he had lived in Siam for twenty years. According to a newspaper article, Gaddi spent eight months in London during his youth, studying natural history specimens in museums. His age is never given, but since 1913 he had been employed steadily as a collector of natural history specimens, first for the Malayan government, then for the government of Siam. From 1926 to 1934 Gaddi worked as a collector and taxidermist for Hugh Smith, the former U.S. Fish Commissioner and head of the Siamese Bureau of Fisheries, who recommended him to Mann. In 1932 Gaddi served as guide to Crown Prince Leopold of Belgium, and traveled with him to the Philippines, Celebes, Bali, and Borneo. Gaddi also collected for the British Museum and for the Philadelphia Academy of Natural Sciences. Mann was not the only one who wanted to hire Gaddi in 1937. Upon arriving in Siam, the Harvard primatologist Harold Coolidge sent for him, but was disappointed when Gaddi chose to stay with Mann. Gaddi may have been exceptionally well qualified, but as a native naturalist earning a living by working for Westerners, he was not unique. Coolidge was able to hire another local taxidermist and collector, Y. Siah, who had previously worked for American scientific expeditions. Nor were experienced native naturalists a recent phenomenon. Eighty years earlier, Alfred Russel Wallace had employed and depended on his Malay assistant Ali.[40]

Gaddi was an important cultural interpreter for the expedition. He spoke the local dialects and served as a liaison between the expedition and the community. When people brought animals to the camp to sell, Gaddi translated. He also was trusted to buy food and other supplies for the animals. In addition, his duties included caring for the animals, as did the keepers Jennier and Davis, and if a specimen died Gaddi was called on to prepare it as a taxidermy specimen.[41]

Tourism

With things in control in their camp and a steady stream of animals coming in, the Manns traveled. On land they drove—roads were not only well maintained, but posted with signs for tourists, marking natu-

FIGURE 19. Gaddi poses with a gibbon and a tiger cub. Smithsonian
Institution Archives. Record Unit 7293. William M. Mann and Lucile
Quarry Mann Papers, circa 1885-1981.

ral wonders and curiosities like places where the road crossed the
equator. The element of tourism in the Mann's expedition was not lost
on their popular audience. One high-school student who read a *National Geographic* article about the expedition thought it sounded like
an ideal vacation, and wrote to Mann as part of a school assignment:
"My classmates and myself are planning 'ideal' vacations that we
would like to take. Upon reading your article . . . about the round-the-
world trip you and an expedition took in search of animals, I decided
that my imaginary vacation would be something like this."[42]

The Manns also traveled thousands of miles by steamship. Every
port seemed to have a small zoo and a collector or two. Coenraad traveled with them and made introductions. The group reached Piroe, on
the island of Ceram, in late April. They stayed a week—long enough
to establish a routine. After they took their morning walk, looking for
ants, Lucile Mann wrote: "The crowd on the street corner, watching for
us, trails us back to the house, all watching to see what we will do

FIGURE 20. As they toured Indonesia collecting animals for the National
Zoo, William and Lucile Mann had their photograph taken near a marker
for the equator. Smithsonian Institution Archives. Record Unit 7293.
William M. Mann and Lucile Quarry Mann Papers, *circa* 1885-1981.

about the one man among them who has a lory or a cuscus, for sale."
As the collection grew, the biggest problem was building enough
cages: "We have two carpenters working like mad trying to keep up
with the specimens coming in. We have bought up all the wire in the
village, and cages have to be devised with wooden bars in front."[43]

After transporting the collection by steamship back to their base
camp, the Manns packed to go home, via Karachi, Ceylon, and the
Suez Canal. Their collection included nearly eighteen hundred specimens accumulated through the various reaches of the network of naturalists, government officials, and animal dealers they had contacted.
Among them were a pair of Himalayan bears donated by Harold Coolidge, a monitor lizard collected for the Manns by the Dutch government, and a Chinese alligator purchased at the last minute from a
dealer in Singapore and transported to the ship in a taxi.

The work was far from over. A delay in loading the animal cages
onto the ship left them sitting in the blistering sun; as a result five hundred birds died. During the trip, even in good weather some cages

FIGURE 21. Zookeepers prepare to feed animals in cages during the fifty-day voyage from Indonesia to the United States. Maynard Williams/National Geographic Society Image Collection.

were hard to reach, packed tightly on the cargo ship; and rough seas sometimes made it impossible to feed and water the animals. Dysentery swept through both crew and cargo. A python got out of its box and killed a cockatoo. Seven out of thirteen gibbons died. By the time they reached Washington, in September 1937, the collection numbered about nine hundred.[44]

The Smithsonian–Firestone Expedition to Liberia, 1940

Three years later, the Manns set out on a collecting trip to Liberia, where their experiences differed in many important ways from those in the East Indies. The Republic of Liberia covers forty-three thousand square miles on the southern coast of West Africa, an area about the size of the state of Ohio. It is bordered on the east by the Ivory Coast, on the northwest by Sierra Leone, and inland by Guinea. Liberia is located slightly north of the equator, in the forest belt of West Africa; heavy rainfall supports dense vegetation and fills dozens of rivers, which flow from the hills of the interior down to the Atlantic Ocean.[45]

Liberia was founded in the 1820s as a colony for the repatriation of freed American slaves, supported by the American Colonization Society, and it became an independent republic in 1847. Americo-Liberians settled along the coast. They became a ruling minority over the sixteen to twenty-eight ethnic groups (estimates vary) living in the interior of the country. In general, settlement of the coast was little contested. The Africans lacked a central political structure, and European traders had passed Liberia by because the region was poor in natural resources such as rubber, palm oil, and ivory, and the coast lacked deep water ports. There were no large slave-trading stations in Liberia. Until Harvey S. Firestone decided in 1923 that Liberia would be a good place for a rubber plantation, the only people to see the region as a place for expansion were Episcopalian missionaries from the United States. The Firestone plantation was the first major foreign investment in the country. By 1946, it employed thirty thousand Africans, recruited from seventeen tribes in the interior.[46]

Western naturalists in the 1920s referred to Liberia as "the least known part of Africa." Firestone sponsored an expedition of Harvard scientists to make a biological and medical survey of the country in 1926 and 1927. Indeed, little was known of Liberia's flora and fauna. The report of the Harvard expedition relied for background material on reports from German explorers in the 1880s, and a 1906 book. Nor was Liberia a hub for zoo collectors, although the ubiquitous Hagenbeck company had been there in 1911.[47]

A few animals from the Firestone plantation had made their way to the National Zoo before the expedition. In 1928, Firestone had presented U.S. President Calvin Coolidge with a pygmy hippopotamus, which was turned over to the zoo. In 1938 a chimpanzee and a leopard had been sent to the zoo from the plantation. Some time after returning from the East Indies, however, William Mann met George Seybold, a rubber plantation manager who had worked in Sumatra. As they swapped travelers' tales, it came out that Seybold was about to embark on a new job in Liberia managing the Firestone Plantation. He invited the Manns to visit. This casual encounter led in 1940 to an animal collecting expedition financed by Harvey Firestone, Jr.[48]

Like the Dutch East Indies Expedition, the trip to Liberia was carried out in the public eye. The Firestone company expected to benefit from newspaper publicity. In addition, some of the collection would be displayed at the Firestone exhibit at the 1939–40 New York World's Fair (the Liberian animals joined the exhibit after the fair had opened). Although Liberia had not inspired a canon of travel literature like the East Indies, in 1940 it could still represent the dark continent of explorers who wrote about other parts of Africa. The one popular travelogue of Liberia emphasized its lack of corruption by Western civilization— Graham Greene's *Journey Without Maps*, published just four years before the Manns' trip. As they had in the East Indies, William and Lucile Mann carried the flags of the Explorers Club and the Society of Woman Geographers. The National Geographic Society supplied a movie camera and color film.[49]

The expedition staff was similar to that of the East Indies trip: William Mann was the leader, and Lucile Mann was secretary, bookkeeper, and nurse to ailing animals. Two keepers from the zoo accompanied them. As headquarters, they had the use of the Firestone plantation, called Harbel, located southeast of Monrovia. A rice shed was converted into a house for animal cages. As guests of the plantation manager, the Manns were introduced to white Liberian society—mainly planters, missionaries, and German businessmen—and to government officials. Taking part in plantation life had a long-term benefit. The Manns' friends at Harbel continued to send animals to the zoo after the expedition was over.

Collecting in Liberia, however, was a much different experience from collecting in the East Indies. Transportation and living conditions were primitive in comparison. Monrovia, for example, did not have a pier. People and cargo had to be transported to shore in a launch. The few roads in the country ran up and down the coast. Inland, people traveled in dugout canoes on the many rivers, or by foot on paths through the bush. The only drawback to walking, according to Lucile Mann, was the bridges, which terrified her, and "which happen along every fifteen minutes. They are usually two limber saplings, tied

FIGURE 22. Porters carried William Mann as they forded a river in
Liberia. Smithsonian Institution Archives. Record Unit 7293. William M.
Mann and Lucile Quarry Mann Papers, *circa* 1885-1981.

together with rattan, and require a better head and a steadier equilib-
rium than I possess." No tourist industry had smoothed the way for
collectors in Liberia.[50]

About a week after their arrival, the Manns set out on their first trip
into the bush, accompanied by one of the plantation managers and his
wife, and eighty porters. They used no pack animals. The porters carried fifty pounds each of "bundles, trunks, and boxes" on their heads,
which included kerosene lanterns, a tin bathtub, plenty of canned food,
and Lucile Mann's typewriter. Much of the time, porters also carried
the Manns, in hammocks fastened to wooden frames. There would be
villages to stay in, but, unlike Sumatra, few government rest houses.[51]

The Manns did not need permits to collect in Liberia, but they faced
different challenges. They soon found that Liberia lacked a culture of
collecting. The usual method of sending out word and waiting for native people to collect animals and bring them to camp did not seem

to work. As William Mann explained to the assistant secretary of the Smithsonian, Alexander Wetmore, "Liberia is the most difficult nut I have ever tried to crack. There are absolutely no native collectors here. Animals as food are worth more than anything we can give them." In the villages the Manns saw few pets, and the Liberians did not keep livestock. To obtain meat, the Liberians went trapping (few people owned guns in the interior) and fishing. In general, food was less abundant than in the East Indies. Lucile Mann wrote: "We find living off the country precarious, and are glad we brought a plentiful supply of canned goods. One day we can buy 15 eggs, another day two; an occasional rather green plantain comes in, but bananas, pineapples, paw-paws, and limes are scarce. Even palm nuts and country pepper are hard to get."[52]

The Manns complained bitterly that Liberians did not know how to care for animals much less catch them; at one point, a porter threw down a cage containing a deer, killing the animal. Although the Manns did not put it this way, part of what made collecting difficult was that the people they depended on as collectors were not familiar with the business. The infrastructure of the animal-dealing business had not reached Liberia. During four months in the country the only collectors they ran into were a German and a Dr. Tate from the American Museum of Natural History. Liberia was remote; missionaries had made converts there, but traders had not.[53]

In an effort to recruit collectors, the Manns joined what was known as a "secret society" of snake handlers. With a couple from the plantation and an entourage of porters, they traveled to a village called Belleyella, in the northern interior of the country. The Manns were told there that if they wanted people to collect snakes for them, they would have to join the local secret snake society. The fauna of Liberia includes many poisonous snakes, and members of the snake society were trained in combining ritual with medicinal herbs to heal snake bites, and to stun snakes so they could be handled.

According to Lucile Mann's account of this adventure, the Manns joined the snake society by paying an initiation fee of eight shillings each, which was distributed among the society members. They were then led, at night, to a dimly lighted hut where they took an oath of secrecy and learned passwords and a handshake. They were shown fetishes and taught society rituals. As part of the induction ritual, snake society members gave Lucile Mann the name "Yangwah," which conferred special powers on her. She explained, it "means that I am the judge, the final arbiter in all discussions; I have the power to 'cut the palaver.'" The local people did not speak English, however, and the way the communication gap was bridged suggests the kinds of problems the Manns must have had in other situations: "We had taken Bobor [one of their most trusted porters] with us as interpreter," Lucile Mann wrote. "Each lesson was given in Mano, translated into Belle, then into Kpesse, then into pidgin English, and often into an English that we could understand, Si [the plantation manager] repeating what Bobor said." To their disappointment, the only animal the Manns gained from their adventure was the snake society's mascot, a Gabun viper. Society members later failed to do any collecting for the Manns.[54]

In this situation and others, the Manns needed a culture broker. Liberians in the interior of the country were not familiar with animal collectors. The whites they knew, for the most part, worked for Firestone or were missionaries. Efforts to explain their activities led the Manns into a who's-on-first style of comedy:

> We hear amusing stories of what our boys think of the leader of the expedition. "What is his name?" they asked, and were told "Dr. Mann." "But what was his name before he was a doctor?" And as to the announced object of the expedition, of course they would not be stupid enough to believe that anyone would really come all the way from America to collect animals from Liberia. They have finally decided that he is out of a job, and just traveling around the country waiting for Firestone to take him on.[55]

Later, a district commissioner, an appointee of the Liberian government, told the Manns that they had been suspected of being missionaries, and the people didn't want to have anything to do with them.[56]

Nonetheless, people did bring the Manns animals, although not as many as they hoped for. The district commissioner proved capable of motivating collectors. When he sent out orders, the Manns received fish, snakes, porcupines, a squirrel, and a pangolin. The District Commissioner later sent them a monkey-eating eagle and a four-hundred-pound pygmy hippo, which was carried out of the jungle in relays by porters. In addition to live animals, the Manns also collected insects, reptiles, amphibians, and fish, which they preserved, for the Smithsonian's National Museum. Two new species were later named after Harvey Firestone, Jr.[57]

Being in Liberia posed problems for feeding and caring for the live animal collection. A sick chimpanzee would take milk only from a syringe. Hornbills were fed by hand and could be fussy: some seemed insatiable, swallowing all the fruit, hard boiled eggs, and palm nuts put in their beaks; others refused to eat at all. The animal collection also had to be protected from driver ants. One night swarms of the ants attacked and killed two antelopes. According to Lucile Mann, the zookeepers "fought them with fire and blow torches," getting badly stung in the process.[58]

Getting the animals home involved the usual logistical problems of shipping schedules and buying and budgeting food. The U.S. Department of Agriculture needed to know their arrival date so the cargo could be inspected and the appropriate animals quarantined. The Manns were fortunate to have calm weather for the voyage, and only three of the larger animals—a deer and two civet cats—died. Their food supplies spoiled, however, and had to be replenished in Dakar. Every day Lucile Mann cut up four buckets full of potatoes for the hippos and antelope. The Manns returned from their nearly six-month expedition on August 7, 1940, with a cargo of about one hundred Liberian mammals, birds, and reptiles. This was in addition to 135 animals sent ahead to the New York World's Fair.[59]

William Mann's expedition to Liberia was probably the last large-scale collecting trip of its kind. After World War II neither zoo directors nor commercial dealers embarked on months-long expeditions to gather thousands of animals. In the field, they were more likely to chase down wildlife on the African plains in a jeep (like Clark Gable in the 1953 film *Mogambo*) than to collect with techniques like drives and pitfalls.[60] With tranquilizer darts, beginning in the late 1950s, collectors attempted to reduce the stress to wild animals of capture and transport, and thus reduce the incidence of injury and death en route to the zoo. Airplane transportation, so much faster than sea voyages, also enabled collectors to focus on shipping smaller numbers of animals with fewer losses.

At Expedition's End

William Mann's reputation as an entomologist and his affiliation with the Smithsonian Institution gave his field trips to collect zoo animals credibility as scientific expeditions among both scientists and the public. To a limited extent, reports of these expeditions traveled through the usual channels of recognition by the scientific community. The expeditions to the East Indies and to Liberia were announced in the journal *Science*. Mann published accounts of all three trips in the *Explorations and Field-Work of the Smithsonian Institution* pamphlet series (he coauthored the one on Liberia with Lucile Mann). Museum curators who worked up the dead specimens Mann brought home published papers on those collections.[61]

More important to the mission of the zoo, however, reports of Mann's expeditions reached wide popular audiences. These reports formed a central product of the expeditions; in addition to collecting animals, the expeditions collected stories. Mann, in fact, did not collect the giraffes and rhinos that were supposed to be the main objects of the expeditions, although he did purchase giraffes. It did not matter, however, because he could tell stories of his failed attempts. Going on an expedition to collect animals was an adventure, and the substance of this adventure was field work. Newsreels, newspapers, and radio broadcasts brought reports of day-to-day expedition progress to popular audiences. After the expeditions, magazine articles and lectures by the Manns and others summed up the experiences of traveling to and working in a foreign country. These stories often focused on the physical details of work—how to set up a base camp, prepare meals for hungry tigers, and transport a pygmy hippo through the jungle.

Expedition work, however, was as much about encountering and navigating through foreign cultures as it was about collecting animals. Popular accounts of Mann's expeditions highlighted the opportunity to meet, work with, or observe exotic peoples as an interesting and important aspect of expedition work. *National Geographic*'s photographer in the East Indies, for example, made extra efforts to report on people as well as animals, and sometimes manipulated his subjects in order to accentuate their exoticism. In Sumatra, for example, members of a native Batak village had performed a dance for the expedition members one evening; the photographer, Williams, requested a restaging in daylight so it could be filmed.[62]

Not surprisingly, accounts of non-Western people tended to fit patronizing and racist stereotypes that were familiar to the American public. Williams, for example, transformed Gaddi from expedition assistant to exotic specimen. Although Mann described Gaddi as "a diminutive fellow, mild mannered, [who] looks like a very much tanned college professor,"[63] Williams assured his editor that when the expedition arrived home, Gaddi would appear suitably foreign: "Layang Gaddi the Dyak ... will wear native dress ... when he greets the reporters and will lay aside his gold-rimmed spectacles the better to look (like a head hunter), if not the better to see."[64] In addition, as a journalist trained to look for stories, Williams tended to see the animals as personalities rather than natural history specimens. One tiger, he wrote, "holds his tooth punches like a movie pug," and "every monkey, lemur, felis minuta, flying squirrel, martin, tree kangaroo, and other specimen will have some quality of individuality."[65]

The story told most frequently about the Manns' expedition to Liberia was about their induction into the snake society. "The Jungle Gave Her the Last Word," announced one newspaper headline, referring to the powers of arbitration conferred on Lucile Mann during the ceremony. The Manns liked to tell the story themselves, too, and related it to President Franklin D. Roosevelt one evening when they had been invited to the White House to show their film of Liberia. When they got to the part about Lucile's new powers, Roosevelt turned to her and asked if she would like to be appointed to the Senate for a few days to see if she could end a filibuster. This made the story all the richer, and the Manns added their Roosevelt anecdote to it in subsequent retellings.[66]

The media did not portray the snake society induction as an effort to recruit native animal collectors. Like the Smithsonian's own press releases, magazines and newspapers related the story as a dangerous encounter with primitive jungle people. The Smithsonian Institution told this tale: "Carried for 5 days through the jungle in hammocks swung from the heads of native boys, they came at last to a jungle village. They were received in awe-inspiring silence by the native medicine men, who obviously resented their presence. . . . After a solemn consultation they gave the American explorer and his wife a formal invitation to become members of their own dread cult."[67]

Encountering the strange customs of primitive people during an expedition was considered part of the adventure of field work. Collecting animals through snake society members was not the only purpose for undergoing initiation. By taking part in the ritual, the Manns brought home something just as valuable as zoo animals to the practice of popular science: a great story.[68]

The expeditions to collect animals for the National Zoo also illustrate the practice of field work more generally: the logistics of carrying out such trips, the kinds of work that needed to be done, and the networks of people that made collecting expeditions possible. Most important, they show how field work was a social endeavor. In Tanganyika and in the East Indies Mann stepped into an infrastructure that made his job easy. An animal trade already existed. Both colonial officials, who gave permission to collect, and native people, whose skills and cooperation made collecting possible, understood the value of wildlife as a commodity to Western collectors. A network of commercial animal dealers, both Westerners and natives, ensured a steady supply of animals in local markets. Experienced culture brokers like Gaddi facilitated transactions. The dealers and culture brokers performed work vital to the success of the expedition, and that was often acknowledged in popular accounts. The prestige of Mann's patrons and his previous field experience also facilitated the expeditions. In addition, colonial bureaucracy provided a physical infrastructure—roads, scheduled steamship traffic, and rest houses—that made travel easy.

William Mann expected that, wherever he went, people would understand at least the business of capturing live animals, if not any scientific justification. But Liberia lacked familiar social and physical infrastructures, and the skills and knowledge of the Liberians were more difficult to assess and employ. Although they were experienced hunters, the Liberians had little knowledge of how to catch animals alive. There were no zoos in Liberia, and the Liberians had no experience as safari guides to Western sportsmen. The Liberians also kept fewer domesticated animals than native people of the East Indies. For Mann's part, his link to the Firestone company gave him the resources—social contacts, a place to work, personnel to hire, and an introduction to the government bureaucracy—that allowed him to collect animals. Yet the Manns seemed oblivious to sources of cultural tension in Liberia that hindered their work: the relationships between missionaries and Africans, and, possibly, the relationships between the Firestone company and its employees and the government, and Liberia's role in the evolving war.

The process of collecting animals and bringing them to the zoo for display was not visible to zoo visitors from the exhibits themselves. But it was brought to public attention by William Mann's expeditions. His animal collecting expeditions encouraged vicarious public participation in the scientific activity of field work. The popular science of zoo collecting thus became a way to enlist public support for the zoo, and was part of constructing the zoo's meaning for its audience. Zoo expeditions represented an approach to zoology that was similar to the National Geographic Society's approach to geography; they encouraged a nonspecialist appreciation of the diversity of nature, and a paternalistic tolerance of the diversity of culture, through immediate, firsthand experience.[69]

CHAPTER 5
NATURAL SETTINGS

The summer of 1911 was exceptionally hot and dry in New York City. In the Bronx Zoo's bison enclosure, according to the zoo *Bulletin*, "The short grass has turned brown, and the buffaloes have established a series of dust wallows. To look out over this rolling plain in its present condition is to immediately recall the prairie country of the great West." The zoo's director, William T. Hornaday, took great pride in his bison display. It was a self-conscious departure from exhibiting animals in cages with bars, or in barnyardlike enclosures. In Hornaday's view, "the larger wild animals of North America . . . should be shown not in paddocks, but in the free range of large enclosures, in which forests, rocks, and natural features of the landscape will give the people an impression of the life, habits and native surroundings of these different types." Hornaday intended the bison display to create for zoo visitors the illusion of being transported to the Great Plains, and to show them the point of view of a field naturalist observing the animals in their native landscape. To ensure a clear view of the animals, Hornaday had searched for the least obtrusive fencing material he could find. The impression of freedom created by the spacious enclosure and unobstructed view of the buffalo—the perceived authenticity of "a prairie scene in New York"—also benefited the animals, by encouraging typical behavior like wallowing in the dust. The bison, the *Bulletin* reported, seemed "really to enjoy the dry and parched condition of their 'stamping-ground.'"[1]

Hornaday believed that the best way to display wildlife was in a "natural setting," and zoo directors from the late nineteenth century to the present have reiterated this belief. Although the phrase has a familiar ring, zoo planners one hundred years ago knew little about the natural lives of most of the animals they displayed, and they did not attempt to demonstrate ecological relationships between animals and their environments. What turn-of-the-twentieth-century zoo planners sought to accomplish with their efforts to "create impressions suggestive of delightful strolls through a natureland"[2] was not to show zoo visitors the habitats of particular animals. Rather, a natural setting was one that approximated an aesthetic ideal, and that evoked a set of emotional responses that middle-class Americans—through tourism, and popular painting and nature writing—could associate with encounters with nature. As a technique that heightened the drama of zoo visitors' encounters with nature, the natural setting was indebted to dioramas of taxidermied animals that were popular at natural history museums. In fact, the men who designed the first "natural settings" in zoos were expert taxidermists who had trained in museums. In creating natural settings for zoo displays, zoo planners interpreted the natural world according to well-established visual and literary conventions. Natural settings in zoos were intended to confirm visitor expectations of a transcendent experience in the presence of natural wonders, to enhance local pride, and to correct popular misconceptions about animals. They served zoos' goal of educating their audiences to the proper appreciation of wildlife.

American zoos originated in a context of increasingly widespread appreciation for the natural wonders of the American landscape. Travelers, artists, and naturalists, with both distinct and overlapping interests and goals, contributed to this heightened interest in American nature. One way this appreciation was expressed in the mid-nineteenth century was through tourism to natural attractions such as Niagara Falls. Wealthy travelers sought the sublime in these sights—including also the Hudson River Valley, the Catskills, Lake George, and the White Mountains. Such tourist attractions were "sacred places"; they were points of pilgrimage important in fashioning a national identity. Lacking Europe's history, Americans found appropriately significant cultural monuments in their landscape.[3]

Later in the nineteenth century the American landscape, particularly the West, attained widely recognized significance as a national symbol. Artists, naturalists, and cartographers accompanied government-sponsored expeditions to explore those territories, and their depictions of wildlife and landscape became accessible to a broad public. The landscape of Yellowstone became an icon of American nature even before it was made into the country's first National Park in 1872—a park set aside mainly for tourism. And tourism became accessible to a broader public—middle-class clerks and bureaucrats who used their vacations to escape to nature. With this greater popular appreciation of nature, and with the domestication of sublime sights through guide books and magazine articles, wonders such as the Grand Canyon may have lost some of their power of transcendence. Tourists still sought scenery, however, and picturesque views retained a capacity for moral uplift, as well as a set of aesthetic values. In the natural world people sought beauty, wonder, and authenticity.[4]

As nature tourism became increasingly popular, urbanites began to seek rejuvenating breaks in nature close to home, for example in public parks. Park and zoo planners hoped to evoke the expected aesthetic responses in their visitors without burdening them with the travail and expense of a train ride or long breaks from work. Parks and zoos were tourist attractions on a local scale. In the words of William Hornaday, the zoo brought "the beauties and wonders of nature within reach of hundreds and thousands who are unable to travel."[5]

To create displays that would evoke appropriate responses to nature, zoos turned to models of visual representation in the work of artists, naturalists, hunters, taxidermists, and people who practiced various combinations of these crafts. American landscape painting provided one example. In the mid-nineteenth century artists such as Thomas Cole and Albert Bierstadt portrayed American landscapes with transcendental beauty. At about the same time, a genre of painting on an even grander scale became popular: panoramas of American scenes covering as much as 45,000 square feet of canvas, which were stored on rollers and unfurled to audiences in theaters over the course of several hours. As travelogue and natural history documentary, such panoramas offered middle-class audiences (who could afford to pay 25 to 50 cents per viewing) another means of appreciating the American landscape.

The panoramas were seen as educational, as well as entertaining, to viewers who would likely never travel to the places portrayed.[6]

Some of these panoramas provided details on the wildlife in natural landscapes. Woodward's Gardens in San Francisco, for example, offered in 1880 "The Zoographicon, or, Rotating Tableaux of Natural History," showing scenes of North America, as well as South America, the Arctic, South Pacific islands, Australia, Africa, Europe, and Asia. According to a guidebook the "animals, birds, reptiles, etc." of each region "are all geographically distributed in the diagrams or in the country they represent, and assume life-like attitudes and movements."[7]

The panorama at Woodward's Gardens belonged to a style of natural historical illustration that emphasized narrative—the story of animals' lives and habits in their native habitat. Illustrations in popular works of natural history also showed animals going about their daily business. During the nineteenth century, taxidermists preparing specimens for educational purposes began to work from a similar narrative perspective, placing their specimens in settings suggesting animal habitats. In the late 1860s, for example, Frederic Webster set his stuffed birds before painted backgrounds and photographed them to produce what he called "animated nature sets" of stereoscopic pictures, which he sold to schools. Webster later was among the first museum taxidermists to mount specimens in groups rather than individually, and with paintings and props to suggest birds' activities in nature.[8]

Taxidermists also were mounting groups of mammals in theatrical poses for display at international expositions in the late nineteenth century. Often violent, like French naturalist Jules Verreaux's depiction of a dromedary being attacked by lions, these dramatized struggle between animals. But the most influential person in promoting a realistic "life group" approach to museum taxidermy was William T. Hornaday, who was a prominent hunter, naturalist, collector, and taxidermist before becoming a zoo director. Hornaday advocated the professionalization of taxidermy. This meant reforming the technique from the equivalent of upholstery to an art form that combined the anatomical knowledge of the naturalist with the aesthetic judgment of the sculptor. According to Hornaday, animals should be mounted in lifelike poses rather than as sportsmen's trophies. A deer, for example, "should have a mild, but wide-awake—not staring—expression, and the attitude should not be unpleasantly strained, either in the curve of the neck or the carriage of the head. Avoid the common error of making a deer's head too 'proud.' "[9] As a museum taxidermist, Hornaday invented technical means of accomplishing these effects. He also set his specimens in what he called "artistic groups"—several animals of the same species, but of varying age and sex, arranged in a natural setting. The accessories that formed the natural setting created a general impression rather than reconstructed a specific location. Hornaday believed it best "to select from a given locality such material as will best represent *an ideal section of the country to be represented as the habitat of the group*" (italics in original). Hornaday created an idealized view of nature with his life groups.[10]

Hornaday's first effort in creating life groups was his "American Bison Group," unveiled at the Smithsonian's U.S. National Museum in 1888. It consisted of six bison specimens representing stages of development from calf to adult female and trophy bull. The animals were not set side by side for comparison of their anatomy, but rather posed in a scene suggesting their wild behavior. The six specimens occupied a glass and mahogany case measuring sixteen feet long, eleven feet high, and twelve feet deep, accessorized with sod, sage brush, buffalo grass, broom sedge, and prickly pear. Hornaday was such a skilled taxidermist that the animals appeared frozen in the moment the bison first caught wind of a naturalist observing them. A reviewer commented: "It is as though a little group of buffalo that have come to drink at a pool had been suddenly struck by some magic spell, each in a natural attitude. . . . The young cow is snuffing at a bunch of tall grass; the old bull and cow are turning their heads . . . as if alarmed by something approaching."[11] Hornaday carried the museum taxidermist's aesthetic of realism with him when he became a zoo director. As chief taxidermist at the U.S. National Museum, he oversaw the Department of Living Animals, a collection that served as models for the museum taxidermists and later became the core of the National Zoo's collection. When Hornaday became the first director of the National Zoo in 1889, and later as first director of the Bronx Zoo in 1896, he attempted to translate the moment-frozen-in-time by a museum diorama to the display of live animals. What could be more lifelike than live animals? Zoo visitors should find animals alert and active, their fur and feathers well groomed. Accoutrements such as grass and leaves, as well as painted backgrounds, which added to the beauty of museum groups, could also be used in zoo displays. Like museum specimens, zoo animals should evoke wonder—an appreciation for the beauty of wildlife. Reforming zoo display methods, by exhibiting animals in large enclosures without bars, helped create an illusion that the animals were free, just as a naturalist would find them in the field.

Ironically, live animals proved more difficult to display naturally than stuffed ones. A natural setting required the illusion that the animals were unconfined, yet the fences containing them were obvious to any viewer. In addition, Hornaday quickly discovered that all natural settings are not equivalent. In New York, soon after he turned loose his first bison specimens on twenty acres of Bronx grassland, they all developed intestinal disease and died. Although perfectly natural, Bronx vegetation did not suit the bison. Furthermore, live animals often did not conform to Hornaday's aesthetic vision of wildlife. Unlike the carefully selected specimens of a museum diorama that were frozen in time, zoo animals shed their fur, defecated, got sick and old, and injured themselves. Zoo animals were difficult to control; national symbols like the bison looked embarrassingly shaggy when they shed their coats in warm weather. Hornaday feared that zoo visitors might find the animals pathetic. He thus sought to control the public view of wildlife by means beyond display in a natural setting. One strategy was to promote the zoo as a studio for artists who would develop a new school of American wildlife painting.[12]

Controlling the View

The usefulness of zoo animals as models for the drawings of artists and naturalists was a common justification for establishing a new zoo. Animal art also served the educational goals of zoos. Natural history illustration had a long (although complicated) association with professional zoology, and drawing the animals was a part of the nature-study curriculum followed by schoolchildren who came to the zoo. Ernest Thompson Seton, an American artist who painted animals and wrote popular animal stories, had used zoo animals as models during his days as an art student in Europe. When he returned to the United

States in the 1890s, he urged the New York Zoological Society to consider the needs of artists. According to Seton, artists required easy access to the zoo and appropriate lighting for their animal subjects, as
well as cooperation from keepers who "only too often are allowed to
regard him [the artist] as an interloper who must keep things pleasant
by continual tipping." At the Bronx Zoo he found an ideal environment for painting.[13]

The Bronx Zoo made exceptional efforts at the turn of the century
to encourage artists to use the zoo's animals as models, and especially
to promote the painting of American big game. After consulting five
artists, including Seton, the zoo constructed a studio in the lion house
with a "wide-screened cage" into which animal models were led, the
first being a big-maned male lion. It opened in 1903. In addition to
providing the studio space, Hornaday arranged a special rate for room
and board at a nearby hotel for artists working in the zoo, and he ordered the zoo staff to be accommodating to artists and sculptors,
whether students or professionals. Under Hornaday's direction, the
New York Zoological Society also collected animal paintings for a gal-
lery in its administration building.[14]

In particular, Hornaday nurtured the career of a young German
painter named Carl Rungius. Rungius, who later in his career painted
diorama backgrounds at the American Museum of Natural History,
shared Hornaday's views about the aesthetics of representing animals.
Like those in museum displays, animals in paintings should be ideal
specimens—healthy males, like hunters' trophies—set in an appropriate landscape. Rungius described the work of the artist-naturalist:
"Whenever the artist intends to standardize a certain species, he must
choose for the landscape that season of the year which will bring out
the characteristic points of his subject. The pelage must be neither too
long nor too short; and the animal must be in good condition." While
observing bighorn sheep in the field, he noticed that in summer, just
after shedding, the animals "wore a short, glossy coat, which brought
out every muscle." By late October, however, "the whole anatomy
seemed altered" by the growth of the winter coat. In Rungius' judgment the summer coat "lacked character" and the winter coat made
the animals look smaller and obscured their lines. Based on these ob-

FIGURE 23. "The Mountaineers: On Wilcox Pass," Alberta, by Carl Rungius. © Wildlife Conservation Society, headquartered at the Bronx Zoo.

servations, Rungius decided to paint the rams in September (in a setting that reflected the season): "Then he certainly is an inspiring subject for the animal sculptor and painter." As Hornaday had done with his life groups, Rungius froze the animals he painted in an ideal setting and season.[15]

Hornaday trusted artists, both professional and amateur, to portray animals in a way that would emphasize their beauty and noble character. He did not have the same faith in photographers, and he could not exercise the same control over amateur photographers in the zoo. Soon after the zoo opened in 1899, Hornaday forbade visitors from carrying cameras into the zoo, and announced that no permits to make photographs could be purchased. The policy stood for forty years.

Hornaday offered the explanation that the zoo had sole privilege to sell photographs of the park, and wished to profit from postcard sales.

FIGURE 24. An Indian leopard was photographed in a cage specially arranged for this purpose, with painted background. © Wildlife Conservation Society, headquartered at the Bronx Zoo.

He also noted that European zoos restricted photography and that the zoo was often too crowded to set up cameras with tripods. But Hornaday objected to amateur photography mainly because he wanted to control the images and memories of animals that people brought home from the zoo.[16] "No photographer can obtain good pictures by making exposures from the walks, or between the bars of enclosures," wrote Hornaday. Such snapshots would include elements of the animal's enclosure—the animals would appear imprisoned. "Good pictures are hard to win," Hornaday continued, "and poor pictures are worse than none, for they repel interest instead of attracting it." Poor pictures might inspire pity for the animals rather than appreciation; nor did they serve the needs of the student of natural history.[17]

In Hornaday's view a good photograph showed a zoo animal in a natural setting much the same as a museum diorama, and conveyed similar aesthetic and zoological information. Taking such pictures required a vision that Hornaday doubted zoo visitors were capable of expressing through photography, as well as technical means to which zoo visitors did not have access. In some cases, the zoo provided its official photographer, Elwin R. Sanborn, with backdrops of painted landscapes to hang behind the animal being photographed. In addition, for good pictures the photographer needed to enter an animal's cage, which visitors obviously could not do. Hornaday elaborated:

> To secure a good picture of a wild animal, the creature must be made to pose! ... The animal must be compelled to halt in the right spot, face in the right direction, and stand fast without nervousness while the camera-man approaches dangerously near, and secures on his ground-glass an image that is something more than a suggestion of a deer, antelope, bear, wolf, or whatever the particular Risk may be.... The ideal picture should show a full side view, with head erect and properly posed; and such accessories as legs, feet, tail, ears, antlers or horns, require adequate representation. A muff-like ball of fur is not necessarily an animal, even though the camera has been brought to bear upon the vital spark.[18]

Photographing animals, in Hornaday's view, required a level of interpretation, and control of the scene, not possible for the average visitor. Souvenir photographs sold by the Bronx Zoo typically showed a single animal, or a group of two or three animals, with the animals filling the frame, their enclosure minimized or eliminated altogether. These photographs aimed to portray the same illusion produced in natural history museum dioramas and the zoo displays themselves, of an animal in its natural setting.[19]

Maintaining the illusion of an animal unconstrained in a natural setting was difficult, if not impossible, to achieve in zoo displays themselves. As the Bronx Zoo's exasperated landscaper put it, "many an ugly corner could be hidden by shrubbery if the animals would only refrain from eating such planting." Most animals remained behind steel bars or fences. And nothing detracted from the aesthetic experience of viewing animals in a zoo more than bars. Barred cages—even large ones—squelched any perception that the captive animal had some degree of freedom. Rather, bars conveyed imprisonment, and gave rise to pity rather than admiration in zoo visitors to whom the opportunity to observe wildlife in some semblance of freedom was important. "Who does not glory in the free, graceful movements of animals, especially wild ones?" asked one zoo critic. "Therefore, this plea for more commodious quarters for our little brothers of the fields, — quarters in which they may have free play and at least a few of the advantages of their natural habitats. This would also afford us all pleasure in place of regret which many of us feel after visiting a zoo."[20]

Beyond concern for the humane treatment of animals, zoo professionals sought to satisfy visitors' intellectual curiosity. "There has been a trend of late to delve more deeply into the secrets of nature," wrote Richard A. Addison, mammals curator at the San Diego Zoo. "Proof of this lies in the consumption of great quantities of nature literature, both fictional and scientific." Addison believed that well-read zoo audiences would appreciate naturalistic exhibits. "The animal's home and natural surroundings are as interesting to most people as the animal itself, and to display it in an old style cage is realizing on only half its value."[21]

Restricting photography, commissioning art, and being aware of zoo visitors' reading habits, however, did not address the zoo director's predicament of controlling the physical space in which zoo visitors observed animals. Balancing the need to keep animals captive while maintaining an illusion that they lived in natural surroundings was perhaps the most important problem facing zoo planners at the turn of the century. A novel solution came from Germany.

Carl Hagenbeck's Panoramas

By all accounts, the German animal dealer Carl Hagenbeck deserves credit as the first showman to display animals confined within moats rather than behind bars or fences. In the 1890s, Hagenbeck began building such displays, which he called panoramas, and he traveled with them to expositions in Berlin, Dresden, Paris, and St. Louis. In 1907, he opened a permanent exhibit—Hagenbeck's Tierpark near Hamburg—a zoo, animal holding station, and showcase for his (by then) patented idea for animal display. Hagenbeck's exhibits capitalized on public preference for viewing wild animals in a setting where

Carl Hagenbeck's Tierpark
Stellingen-Hamburg
Nordland-Panorama

FIGURE 25. The arctic panorama at Hagenbeck's Tierpark is shown in this postcard. Library, National Zoological Park, Smithsonian Institution.

they appeared to be free. This desire presumably stemmed from both humane and aesthetic impulses. Aware of complaints about zoo animals kept behind bars, Hagenbeck wrote, "I wished to exhibit [the animals] not as captives, confined within narrow spaces, and looked at between bars, but as free to wander from place to place within as large limits as possible, and with no bars to obstruct the view and serve as a reminder of the captivity." Hagenbeck's business, however, was entertainment, and his panoramas also provided drama; he used moats so that no obvious barriers separated predators from their prey, or from human onlookers. Visitors could experience both intimacy and thrill at their close encounter with wild animals at Hagenbeck's Tierpark.[22]

Hagenbeck's panoramas pulled together elements of popular entertainment and landscape architecture already in existence. As in the United States, panoramic paintings became popular in Germany in the nineteenth century, as did another type of exhibit referred to as a panorama—diorama scenes that used sculpture and lighting, as well as paintings, to reconstruct historical moments, such as battle scenes. Using moats to confine animals also was not an original idea, rather one adapted from the ditches and sunken fences, known as ha-has, used to contain livestock on eighteenth-century English country estates. Hagenbeck brought these elements together in a new way. He added live animal actors to panoramas of natural scenery, using moats to confine them.

Hagenbeck's first panorama, and the one he showed at the 1904 world's fair in St. Louis, was an "arctic" panorama featuring seals, sea lions, polar bears, and penguins, in a scene constructed to resemble a polar landscape. As in the panoramas built later at his zoo, the animals were displayed on a series of landscaped stages, each rising up higher than the one before it, and separated by moats. In the "animal paradise" at Hagenbeck's zoo, water birds swam in a large lake in the foreground. Beyond them stretched meadows occupied by flamingos, cranes, pelicans, and ibises. Farther back, lions prowled in their grotto, and the scene culminated in a rocky hill where sheep, goats, and antelope could climb. Creating the artificial mountains and gorges of the Tierpark's landscape required moving more than 55,000 cubic yards of soil. The Swiss sculptor Urs Eggenschwyler assisted in designing artificial rocks rendered in concrete. The exhibit could be viewed as a whole from the zoo's central restaurant, a point "fixed in the garden from which might be seen every kind of animal moving about in apparent freedom." Or visitors could walk on paths between the separated displays (densely planted foliage concealed them from viewers sitting in the restaurant). The zoo was an immediate success with the German public.[23]

American zoo directors quickly became familiar with Hagenbeck's displays, both through his fair exhibits and through their ties to his animal-dealing business. Sol Stephan, director of the Cincinnati Zoo, was a Hagenbeck representative in the United States, as was John Benson, briefly director of Boston's Franklin Park Zoo, who later opened his own animal park. Many American zoos purchased animals from Hagenbeck, and American zoo directors always included Hagenbeck's Tierpark in their tours of European zoos.[24]

American zoo directors, however, remained cautious about constructing moated displays in their own zoos. William Hornaday was dead set against them. In his opinion, the educational value of the zoo decreased in direct proportion to the increase in distance between animal and viewer. Hagenbeck tried to persuade Hornaday to build a moated bear display, but Hornaday found that it "would push the animals so far away from visitors that fully one-half their potential educational value would be lost. People who study animals always desire to get as close to them as possible." Furthermore, Hornaday believed that the large and expensive exhibits would require a cut in the number of species a zoo would be able to display. In the 1920s he tried to persuade other zoo directors to protest the "Hagenbeckization" of American zoos. Anti-German sentiment also contributed to Hornaday's objections to Hagenbeck's innovations in animal display.[25]

Other zoo directors debated the merits of moated displays in terms of animal management. In the days before antibiotics, when good sanitation was the main defense against animal disease, zoo directors wanted to know whether barless displays were harder to keep clean than cages. In addition, the Smithsonian's Alexander Wetmore feared that a fallen tree or heavy snowfall would allow animals to cross moats and escape.[26] Joseph Stephan in Cincinnati commented that feeding animals in groups on moated exhibits could be problematic, especially "preventing the strong animals from attacking or stealing from the weak. Certain animals require more food than others, some require special kinds of foods, and it usually is very difficult to see that each animal gets its proper share."[27] Finally, although Hagenbeck advocated acclimatizing tropical animals to cold winters, American zoo directors were not confident that big cats and primates would survive the winter of, say, Milwaukee outdoors. Building moated displays for these animals to inhabit during the summer would thus entail expenses in addition to maintaining their indoor winter quarters.[28]

Nonetheless, the unobstructed view of the animals offered by Hagenbeck's innovation held wide appeal. It was a way of getting rid of imprisoning bars. Just "one or two barless enclosures, in that they are marvelous spectacles, would help make either the small or larger garden have a greater popular appeal," observed Robert Bean, director of the Washington Park Zoo in Milwaukee.[29]

Moated displays also gave zoo planners new options for exhibiting animals in natural settings. American zoo directors quickly began referring to the Hagenbeck park as a "habitat" zoo. A few of Hagenbeck's displays attempted to put animals in a landscape that evoked their native habitat—the Arctic panorama in particular. But this was not uniformly the case. German zoologists criticized Hagenbeck for mixing deer and antelope from all over the world in one display. Although Hagenbeck's park, in retrospect, has been labeled a harbinger of zoogeographically organized zoos, it was not planned as such. When American zoo planners referred to Hagenbeck's zoo as a "habitat" zoo, they meant that the display design put the animals outdoors, in a setting large enough for them to be active in a way perceived as similar to their behavior in nature. Goats, for example, were provided with artificial mountains to climb.[30]

American zoo directors criticized Hagenbeck, however, for the fake look of his natural settings. The concrete imitation rocks were obviously an artist's conception of rocks, modeled by hand. They replicated neither a generic type of rock nor rock from a specific location. "The rock work is obviously cement and of little artistic merit," wrote one critic.[31] Furthermore, noted a different commentator, "no attempt was made to give to the rocks the natural shades, or to embellish the work with foliage,"[32] in Hagenbeck's park. This fakeness detracted from both the exhibit's aesthetic appeal and its educational value, in the view of American zoo planners.

The first Americans to adapt Hagenbeck's idea of displays with moats wanted their natural settings to be more authentic. Rather than generic concrete rocks, they created painstaking replicas of local geological formations. These natural settings bore no relation to the habitat of the animals eventually displayed in them. But they conformed to American pride in natural wonders, and an interpretation of the meaning of the zoo as a local achievement. With neither "the beautiful relics of an ancient civilization nor river banks set with the castles of a romantic age,"[33] in the words of one zoo booster, Americans looked for monuments to their history in nature, both on a national and on a local scale. The first zoos to build moated displays linked the idea of the natural setting to local geology. These displays symbolized on a local scale what monuments like the Yosemite Valley stood for on a national scale. Meticulous copies of local natural wonders in zoos heightened the appeal of zoos as tourist attractions, and popular zoos boosted civic pride. At the same time, this devotion to geological authenticity helped zoos maintain their claim to status as scientific and educational institutions.

Bears and Boosterism

The Denver Zoo began plans to build displays with moats in 1912. The project was the effort of the zoo director, Victor H. Borcherdt, and the city of Denver's landscape architect, Saco R. DeBoer. Borcherdt, a native of Denver, had learned taxidermy from his father and was the locally well-known creator of dioramas for the Denver Museum of Natural History before being hired by the zoo. DeBoer was known as a proponent of landscaping using native plants. Together they proposed rebuilding the Denver Zoo as a "habitat zoo," modeled on Hagenbeck's design. The natural setting they sought to create in the zoo was a mountain ridge in the Colorado Rockies. The original plan called for an artificial mountain of concrete applied to a reinforced steel framework 332 feet long. A 200-foot-long structure was completed in 1918, with space for bears and monkeys.[34]

Because bears could be displayed outdoors all year round, bear displays were a logical place to start in building moated exhibits. In European menageries and zoos since the sixteenth century, bears had been kept in deep pits with a tall pole in the center. When the bears climbed up the pole visitors could see them and feed them. In the nineteenth century some bear enclosures were built at ground level with heavy iron bars, but they were referred to, nonetheless, as the bear pits. A private zoo in Denver had a typical bear pit, whereas the Philadelphia Zoo followed the iron bar model. The Denver Zoo promoted its new bear display as humane animal management—a means of freeing the animal prisoners from "damp pits, in which the sunlight could penetrate only to a little spot on the back wall," and "trying to give back to them, be it in a limited way only, some of the surroundings they were removed from."[35]

To create a Colorado Rockies scene at the zoo, Borcherdt marshaled his experience in mounting stuffed life-group displays. Like Hornaday with his bison display, Borcherdt wanted to create an illusion for zoo visitors. Borcherdt, however, decided to copy a specific place rather than produce a general impression. The place he chose was the summit of Dinosaur Mountain, near Morrison, Colorado, and to make his model he applied a technique of taking plaster casts from natural cliffs that he had developed for his museum work. Borcherdt and his assis-

FIGURE 26. The bear pit at Elitch's Zoological Gardens in Denver, circa 1893, was a typical display. Proprietor Mary Elitch entered the enclosure. Denver Public Library, Western History Department.

tants spent several weeks camping on the mountain peak, first fashioning a mold of the rocks out of rubberlike glue strengthened with cement and hemp fiber, and then making plaster casts from the mold. Burros hauled materials up the mountain, and the casts were transported back down by bobsled. Back at the zoo, workers built a steel frame, to which the cast-cement cliffs were attached.[36]

Authenticity was of utmost importance. DeBoer, the landscaper, reported that, "Not a crack or seam, nor a projection, which was on the natural rock is omitted on the artificial rock. Even the lichens show." Mineral pigments added to the concrete duplicated the color of the Dinosaur Mountain cliffs. When completed, the steel and concrete cliffs measured forty feet high and two hundred feet long, and were surrounded by an eighteen-foot-wide moat. Then DeBoer added the "finishing touch" of vegetation, filling crevices with dirt, and transplanting dozens of shrubs and wildflowers that grew in the natural cliff surroundings. In addition he used more standard landscaping material, but with sensitivity to the overall effect. "Bunches of Kentucky

FIGURE 27. Hauling supplies up the mountain. Denver Public Library, Western History Department.

FIGURE 28. Workers make casts of the cliffs. Denver Public Library, Western History Department.

FIGURE 29. A steel frame for the display was constructed at the zoo. This
end of the exhibit was originally intended for rhesus monkeys. Denver
Public Library, Western History Department.

blue grass were also used at first," DeBoer wrote, "but were later removed as being too green for this landscape, the color of which in nature runs more towards gray and blue than to green."[37]

Authenticity, however, was not extended to the species of animals
displayed on the artificial mountain. Reviews made no mention of
what kinds of bears occupied the exhibit when it opened, although
within a few years it housed sea lions, and it has since been home to
polar bears. The portion of the display built for the monkeys (rhesus
monkeys from India) illustrates again how the exhibit served to link
the natural setting with local tourist attractions. At one end of the display, beneath an enormous overhanging rock formation, the zoo designers built a replica of one of the Mesa Verde cliff dwellings, "complete even to the primitive ladders." A few reviewers commented
with distaste on the analogy between the monkeys and the Native
Americans. But most focused on the architecture. Denver's mayor

FIGURE 30. The completed bear display was a reproduction of Dinosaur Mountain. Denver Public Library, Western History Department.

boasted: "The effect is so realistic that visitors examine the concrete with incredulity, doubting the attendant's statement that the rock is manufactured."[38]

Based on Borcherdt's success in Denver, the St. Louis Zoo hired him in 1919 to build an even more elaborate bear display for the zoo in Forest Park. The St. Louis exhibit was to be built as a series of nine three-sided sections, with the moat in front. Upon completion it would measure one thousand feet long. In St. Louis, however, Borcherdt was not to erect a duplicate section of the Rocky Mountains. Rather, the members of the St. Louis Zoological Board of Control, the zoo's governing body, made several expeditions to choose an example of local geology to re-create—rocks that would symbolize the state of Missouri. The board members finally settled on a section of limestone palisades along the Mississippi River near the town of Herculaneum, Missouri, fifty miles south of St. Louis. Again, unlike museum naturalists, the St. Louis Zoo's planners chose a site that had no connection to the habitat of the animals to be displayed.[39]

The St. Louis Zoo board considered Borcherdt the perfect man to design the display, both for his practical experience and for the respectability his knowledge as a naturalist gave their project: "Not only was it necessary to have a person to build these quarters who knew how to handle cement, sand and water," according to a member of the zoo board, "but he also had to have a knowledge of the nature of the animals to be housed and had to be familiar with the geological structure of a limestone formation."[40] They did not consider the display merely a spectacle; authenticity supported the zoo's educational mission. The St. Louis Zoo's planners also considered the bear display to be humane. They hoped the added space of their habitat displays would discourage the pacing behavior many bears adopted in cramped quarters. The St. Louis design, for example, also included dens behind the rock face to which the bears could retreat, and which remained cool in summer, as well as excavated pits that the designers hoped the bears would use for breeding.[41]

The concrete reproduction in St. Louis was colored and landscaped not just to resemble the original cliffs, but to grow more like them over the years. "Its beauty and naturalness is expected to be enhanced with time as the concrete of which it is constructed grows into a more uniform semblance of the natural limestone from which it was copied," reported a newspaper. Three black bears imported to the zoo from the Pacific Northwest became the first inhabitants of the display. Later an Alaskan "peninsula" bear and polar bears joined them. And, as in Denver, reviewers complimented the realism of the display. The bear display also advanced the reputation of St. Louis, a city particularly self-conscious about attracting tourists and new residents with its civic institutions. When the first section of the display opened in 1921, the zoo received national press coverage and was featured in newsreels. The zoo's director, George Vierheller, claimed the zoo had earned St. Louis international fame.[42]

Riding out this success, the St. Louis Zoo planned additional natural settings for new displays built in the 1920s, mixing the goals of authenticity, natural beauty, local pride, and humane treatment of animals, while divorcing these settings from the natural habitats of the animals displayed. As the bear dens were completed, zoo board members planned another display, a reproduction of the African veldt, intended to house antelope, giraffe, and zebra. Again they scoured the local landscape for a distinctive geological feature to reproduce in concrete. They chose a formation of granite boulders near Graniteville, Missouri.

Like the limestone cliffs in the bear display, the granite boulders helped increase awareness of local geological treasures. The boulders concealed an animal house and were landscaped with native dogwood, redbud, and oak. The zoo selected a similar site as a model for its lion dens. A souvenir book summed up the many meanings of nature addressed by choosing the granite formation, which

> on account of its beautiful coloring, clean, strong lines, will display these animals to their greatest advantage. Lions, tigers, leopards, etc., among the animal kingdom typify strength as does granite in mineral science. . . . Furthermore . . . the duplication of the granite . . . will further the scientific interest of the wonders found in our own state. Thus the cat tribe will be located in naturalistic surroundings and when housed in their new homes, they will no longer be pitied for their imprisonment, but will be admired by the public in their natural environment.[43]

Furthermore, the reddish color of the granite boulders would "have a warm effect in keeping with the natural tropical habitat of the cat animals." Zoo officials acknowledged that this was clearly not their native habitat: "Those who go to the zoo will see the granite ridges and boulders of Iron county as they would look with lions, tigers and leopards wandering about in them." Rather, it was a setting that would inspire suitable awe in the presence of the animals, heightened by their juxtaposition to a local natural wonder.[44]

The zoos in Denver and St. Louis were particularly explicit about tying the construction of natural settings to local pride. But other zoos also copied local geology as a backdrop for animals from other places. In the 1930s, the Cincinnati Zoo planned an African veldt display with a design that would conceal the architecture of the building: "Seeking something from which to copy, motion pictures of the rock formations along the majestic Kentucky River were made. These have been reproduced in concrete on the exterior of the building. Such attention has been devoted to even the most minute details that the entrances to the various stalls and interior enclosures themselves will simply appear as crevices or caves in the rocks." The zoo in Evansville, Indiana, built a barless bear display from limestone, but stained the finished display because the limestone was white, "whereas the local stone outcroppings which we were striving to imitate runs to grays and browns," wrote the director.[45]

American zoo planners in the 1910s and 1920s perceived Carl Hagenbeck's innovation of confining animals within moats as an opportunity to display animals in their "natural setting." Animals in these displays appeared relatively free, compared to those in cages with bars. Using moats cleared the sight line between animal and visitor, allowing both appreciation of the animal's beauty and the thrill of proximity to a wild, and dangerous, creature. What zoo planners meant by a natural setting, however, had little to do with the ecological relationship between the animal displayed and its natural habitat. Rather, they adapted the idea of moated enclosures to American conventions of viewing nature. They constructed concrete cliffs—meticulously planted and painted—in zoo displays as an expression of pride in their city, and evidence of its history, permanence, and importance. The way zoos adopted local geology as important to civic culture was similar to the way in which Americans as a nation adopted natural wonders as national treasures.

The technological achievement of creating a realistic reproduction of nature also contributed to the appeal of these displays. Audiences derived pleasure from the trick, and such local feats of skill further validated civic pride in the zoo. Historian David Nye argues that in American culture the experience of the sublime moved "from awed contemplation of natural scenery to the rapt enthusiasm for technological display," and that, at the same time, the emotion associated with the sublime was transformed. Whereas sublime once referred to a glimpse of divinity in nature, in the twentieth century it became the evocation of nationalism by technology. In zoo exhibits, in a geographically more limited way, the two notions of the sublime combined. The technical skill and the faithfulness to reality with which the illusions were constructed heightened the awe and amazement of viewing wild animals not confined by bars.[46]

Appreciating Reptiles

If an important goal of American zoo displays early in the twentieth century was to evoke customary appreciation of nature's beauty, then reptiles presented a problem. To be sure, reptile displays were popular.

The reptile house was the first permanent structure at the Bronx Zoo, and it was centrally located. The St. Louis Zoo opened a widely admired reptile house in 1927, and other zoos updated their snake displays in the 1920s and 1930s. According to the National Zoo's director, William M. Mann, the reptile house was always the most crowded exhibit at the Bronx and Philadelphia zoos. European zoos capitalized on the popularity of their reptile houses by charging an extra admission fee.[47]

Reptiles were popular for the wrong reasons, however, from the point of view of zoo professionals. People found snakes irresistibly repulsive, ugly, slimy, sinister, and dangerous. As Richard Addison, at the San Diego Zoo, put it, "A very large percent of the people who visit the reptile house do so, I believe, in anticipation of personal thrill, in much the same manner as they read a blood curdling novel." To be fair, zoos catered to public taste. Huge crowds gathered at the San Diego Zoo to watch the 24-foot-long reticulated python Diablo being force fed ground meat with a sausage stuffer. Although gratified by visitors' interest in snakes, Addison continued, "to turn it into the right kind of interest is somewhat of a problem."[48]

Cultivating "the right kind of interest" became the task of zoo reptile houses. To accomplish this, zoo planners adopted a style of display in reptile houses similar to the habitat dioramas that had become popular in American natural history museums: they put the animals in glass-fronted cages, with foliage and a painted backdrop, arranged around the perimeter of an exhibition hall. Lighting the cages and keeping the hall dark, as in museum displays, reduced glare on the glass and focused attention on the animals. Zoos that could afford to do so housed these small-scale natural settings in massive buildings—museumlike buildings that conveyed scientific and cultural authority.

In their efforts to convince their audience that snakes' "patterns and colors, together with their lithe graceful bodies make them a thing of great beauty,"[49] herpetologists at zoos had to rout long-standing prejudices. The "treacherous nature" of the snake "was made famous through his little escapade with Mother Eve and he has enjoyed a bad reputation ever since," according to one zoo man.[50] Whether or not public aversion to snakes had biblical roots, reptile keepers lamented that hikers and campers killed every snake they encountered, indiscriminately. Furthermore, in the 1920s reptile enthusiasts perceived an increase in both snake killing and the incidence of snake bite fatalities,

as campers, tourists, and picnickers ventured by automobile into poisonous snake territory. Teaching people to distinguish poisonous from
nonpoisonous snakes, how to avoid getting bitten, and how to treat
snake bites when they occurred, thus became urgent to the educational
program of zoo herpetologists.[51]

Reptile keepers also attempted to correct a profusion of myths and
fantastic snake stories among zoo visitors. "One of the most widespread misconceptions about snakes is that they can be 'charmed' by
persons with special powers usually Hindus," observed one writer.[52]
In an attempt to answer the commonly asked questions at the Bronx
Zoo, veterinarian Reid Blair stated flatly: "There is no "hoop snake"
that can put its tail in its mouth, stiffen its body into circles and roll
down hill like a barrel-hoop; no "joint snake" that can separate itself
into pieces and later assemble the sections and become a whole snake
again; and no "coach-whip snake" able to lash his assailant with a
stinging tail. There is a milk snake, but since its stomach capacity for
liquids is about three teaspoonfuls, something else must be blamed
when the cow comes from pasture with her udder dry."[53]

Stories about rattlesnakes and their poisonous venom comprised almost a separate mythology. "Will a rattlesnake cross a horsehair rope?"
asked a correspondent to the National Zoo, who hoped to protect himself from being bitten.[54] In the American southwest, the Hopi snake
dance had become a popular tourist attraction. During the performance
dancers carried live rattlesnakes in their mouths, which, according to
San Diego's reptile keeper, prompted onlookers to "comb their dictionaries for new synonyms for terrible, fantastic and repulsive."[55]

Although stories about snakes were most common, zoo professionals refuted myths about other reptiles as well. In 1928, residents of
Eastland, Texas, claimed to have removed a live horned toad from the
31-year-old cornerstone of a building being demolished. Newspapers
quoted William Mann and the Bronx Zoo's Raymond Ditmars, who
said that it was impossible for the animal to have lived so long without
food and water. But, according to newspaper reports, even expert testi-
mony failed to debunk the "West Texas tradition that such reptiles can
hibernate for a century."[56]

Reptile keepers combated superstitions about snakes and other reptiles by various means. Several zoo herpetologists became popular

writers on snakes during the first half of the century. They also developed educational programs that gave zoogoers the opportunity to handle nonpoisonous snakes. Not least, however, they promoted reptile appreciation through a museumlike style of display.[57]

The National Zoo's Reptile House

When William M. Mann became director of the National Zoo in 1925, one of his goals was to erect new animal houses on the zoo grounds. First on his list was a reptile house. The zoo had never exhibited many reptiles for lack of suitable enclosures. Mann decided to request funds for a reptile house before any other structure because, as he put it, the bad public image of reptiles made it "the hardest to get." Indeed, congressional appropriation of funds for the reptile house in 1929 (the usual means of support for the National Zoo at the time) met with some public opposition. A letter to the editor of the *Washington Star* objected to spending money on a state-of-the-art reptile house when the city's schools—in particular seventy-five temporary "portable" school houses—needed repair. While the zoo director planned elaborate heating, lighting, and ventilation schemes for the reptiles, "Children in the portables sit in dark rooms on a cloudy day . . . for there is no artificial light. On cold days they huddle around a stove." Despite the complaints, the reptile house plans went forward.[58]

The National Zoo's reptile house was to incorporate the most up-to-date knowledge on reptile keeping. In the words of the Smithsonian Institution's secretary, it was to be "one of the model buildings of its kind in the world." To this end, the Smithsonian sent Mann on a tour of twenty European zoos in 1929, accompanied by Washington, D.C.'s municipal architect Albert L. Harris. Mann and Harris were directed to study the design and construction of all zoo buildings, but to focus on reptile houses. Although favorably impressed with most of the zoos they visited, Mann and Harris decided to plan their new building after the London Zoo's reptile house, which had been completed in 1927 and itself had been praised as the most sophisticated building of its type.[59]

At the London Zoo's reptile house, the exhibition hall was a large rectangular room, and visitors made a circuit of the building looking at cages arranged around the walls, as well as on a central island. The temperature of the cages was controlled separately from that of the public area so that the snakes could be kept warm, and thus active, while visitors remained comfortable. Light entered the room through skylights over the cages, and the public areas were kept dark, a technique referred to as "aquarium principle" lighting. The reptile house curator landscaped individual enclosures, and a theatrical scene artist painted cage backgrounds. On its second floor, the London Reptile House housed laboratory and office space.[60]

When the National Zoo's Reptile House opened in 1931, it incorporated all of these features. With dimensions of 200 feet by 82 feet, it was also about the same size as the London building. In addition, the glass skylights at the National Zoo's Reptile House permitted ultraviolet light to pass through, which was thought to be beneficial to reptiles. The cages and public space were ventilated separately. And in each glass-fronted cage, "the vegetation, temperature, humidity, and light of its occupants' natural habitat have been carefully simulated," a newspaper reported.[61] Mural backgrounds for the cages had been painted by Bruce Horsfall, a well-known bird illustrator and museum mural painter, as well as another artist, Elie Cheverlange.[62]

More than any other animal house at the National Zoo, the interior of the Reptile House resembled a natural history museum. Thick glass separated visitors from the heat, humidity, and smells of the cages; the murals and plant props created self-contained natural settings similar to museum dioramas. The focused lighting created the same "devotional alcoves" museums provided for proper appreciation of habitat dioramas. In museums, this arrangement was meant to evoke appreciation of a taxidermically prepared animal's beauty. The construction of a reptile house that so closely paralleled a hall of museum dioramas was an effort (whether conscious or not) to raise the status of its maligned residents.[63]

The architect Albert Harris also designed the exterior of the National Zoo's Reptile House to inspire respect for the inhabitants within. A palatial red brick structure, its architecture closely resembled that of Romanesque cathedrals in northern Italy, particularly those in Ancona and Verona. This style enabled Harris to embellish the building with animal sculpture analogous to religious iconography. The columns of

FIGURE 31. An eastern indigo snake on display in the National Zoo's Reptile House, 1934. Photo Archives, National Zoological Park, Smithsonian Institution.

the portico, for example, were supported on the backs of two stylized turtles. In place of saints, crocodiles and toads serve as quoins and ornaments on the Reptile House. Massive carved bronze doors depicted reptiles and their prehistoric predecessors in place of biblical imagery. In the nineteenth and early twentieth centuries, natural history museums were likened to "cathedrals of science"; the design of the National Zoo's Reptile House suggests that it was intended, if not as a place for the worship of reptiles, at least as a place where they would receive deserved attention and respect.[64]

Other zoos also adopted a museum diorama style of displaying reptiles. Marlin Perkins, before he achieved television fame for his nature show, was reptile keeper at the St. Louis Zoo. In his view, this style of display was for reptiles a transformation much like that from barred cages to moated displays for mammals. Referring to the reptile house in St. Louis, he wrote: "To view the exhibit now one sees not a pile of snakes in the corner of a bare cage, but a group of water snakes in much the same attitude and pose as they are seen in the wild state. Some drape themselves over the rim of the old log, others are partially hidden beneath it. . . . To be looking at this exhibit and catch a glimpse of the folds of a snake lying coiled under the stump gives one the impression of discovery." The Bronx Zoo had long used painted panoramic scenes in its reptile exhibits. Cincinnati's reptile house opened in 1937 with "natural habitats . . . accurately reproduced by Public Works Administration artists." The reptile house in Baltimore opened in 1948 with three jungle scenes, two desert, and one cypress swamp, among others—"realistic" settings, in which the foreground blended with background murals.[65]

From Natural Settings to Landscape Immersion

Through the 1940s, 1950s, and 1960s American zoos continued to build displays with moats and to landscape reptile and bird displays with plant props. In many displays, the moat served only as a barrier between visitors and animals, rather than as a boundary between the park and the "natural setting." The aesthetic of realism was often discarded; most zoos had a monkey island, for example, with playground equipment or houses that made the island look like a small Midwest-

MONKEY ISLAND, LINCOLN PARK ZOO, OKLAHOMA CITY, OKLAHOMA

FIGURE 32. The zoo in Oklahoma City built its monkey island to resemble a sunken ship. Undated postcard, author's collection.

ern town. Another popular theme for monkey islands was, in place of a grassy island, a half-sunken ship surrounded by the moat. Other exhibits claimed to be naturalistic, but relied on abstract representations of nature: "Wrinkled concrete came to represent geologic formations, packed clay replaced waving grasses, and a few house plants portrayed the tropical forest," according to one designer's critique. Many of Hagenbeck's innovations—multiple viewing areas for different perspectives on an exhibit, and the arrangment of exhibits so that predators looked as though they had access to prey animals, for example—had been forgotten. The moated display, in any case, was no longer novel. It had lost its capacity to surprise and engage visitors.[66]

In the mid-1970s, the Woodland Park Zoo in Seattle hired the firm of Jones & Jones to design a long-range plan for renovating the zoo. The exhibit designers revived many of the elements of early "natural settings"—a museumlike devotion to realism and authenticity. Their goal was to transport visitors in their imaginations to places in nature beyond the access of most tourists. The gorilla display, for example, recreated a section of upland forest in the part of West Africa that is the animals' habitat. The designers also incorporated ideas from animal behavior, such as flight distance, building into the exhibits places for the animals to retreat as well as devices such as heated rocks to draw them near public viewing areas. The biggest innovation in the design for the Seattle Zoo, however, was to extend the re-created habitat beyond the moat. At the zoo in Seattle, visitors step off the main zoo thoroughfare and into the landscape with the animals before they approach the display—the central concept of "landscape immersion." As the designer John Coe explains the idea, "instead of standing in a familiar city park (known as a zoological garden) and viewing zebra in an African setting, both the zoo visitors and the zebra are in a landscape carefully designed to "feel" like the African savanna." Based on the landscape immersion idea, dozens of realistic tropical rain forests and African savannas were built in American zoos in the 1980s and 1990s.[67]

The goals for these new displays resonate with both the century-old impulse to evoke awe and wonder by placing animals in "natural settings" and the conservation mission of late-twentieth-century zoos. Like zoo directors of the past, exhibit designers today aim for an idealized realism in their displays. They try to re-create places in nature so that visitors feel as if they have been transported to another continent. The effect of this disjuncture, the surprise and the wonder, has long been associated with looking at the natural world; the landscape immersion exhibit is built to provide an aesthetically memorable, enjoyable experience. The perceived lack of barriers in Hagenbeck's displays alarmed the first visitors to see wild animals that appeared to be free, nothing preventing them from charging or attacking; similarly, designers today aim to "make the hair stand up on the back of your neck" in landscape immersion exhibits. Like an encounter with a bear on a hiking trail, a visit to a landscape immersion exhibit is meant to evoke both awe and a pleasurable sort of terror.[68]

Although the landscape immersion exhibits of the last twenty years are superficially similar to earlier "natural settings," what they seek to accomplish by presenting the public with natural spectacles reflects the culture of the turn of the twenty-first century. Rather than copying local geological formations, recent zoo designers have reproduced threatened habitats. In the past, zoo planners hoped to produce a transcendent experience in order to evoke civic pride and an interest in natural history in zoo visitors; today that experience is meant to impress the plight of wildlife on viewers, in the hopes of translating that knowledge into political action, financial support for conservation organizations, or both. The new exhibits are monuments to habitats and animals facing imminent destruction and extinction. The threats to their existence make their accessibility in the zoo that much more spectacular.[69]

CHAPTER 6
ZOOS OLD AND NEW

The ten worst zoos in the United States made headlines on February 19, 1984, in an article in *Parade* magazine, delivered in Sunday newspapers across the country. The article was based on a report by the Humane Society of the United States, and it focused national attention on crumbling facilities and substandard keeping practices at the zoos in Boston, Atlanta, and Brooklyn, as well as zoos in smaller cities. Criticism of Atlanta's zoo, harsh to begin with, became even more strident when further stories of missing and dead animals were revealed. The news came out that the zoo had lost its accreditation with the American Association of Zoological Parks and Aquariums a few months before the *Parade* article, because of its poor medical records, small cages, stressed animals, lack of signage, and dangerous children's zoo. Most appalling was the revelation that the Atlanta Zoo had sent its elephant Twinkles—purchased in the 1970s with pennies, nickels, and dimes donated by children—to a circus where she died.[1]

The *Parade* article and the extended media coverage of problems in Atlanta prompted the public, zoological societies, and city councils across the country to examine their local zoos, bringing national attention to problems that were beginning to be acknowledged in a few cities. Many zoo professionals look on the article as a turning point; it launched what they refer to as a "revolution" in the zoo world. Zoo supporters sum up the milestones of this transformation briefly: in the past, zoos boasted about the number of species they displayed; they exhibited single animals in rows of cement or tile cages with bars; the bored animals paced back and forth or begged for peanuts; and cage labels listed little more than an animal's Latin name. Today, an up-to-date zoo shows fewer species and displays social animals in groups; immersion exhibits replicate animal habitats; and with friends to interact with and toys to help pass the time, animals behave more as they would in nature. In addition, extensive interpretation in the form of signs, videos, and interactive displays educates zoogoers about animal diets, habits, and natural environments, the status of wild populations, and zoo-directed research programs. The overarching educational message is wildlife and habitat conservation. In general, zoos have improved the quality of animal keeping and exhibitry, widened the scope of education, and placed new emphasis on research.

Zoo managers today describe the recent renaissance of American zoos in terms much like those invoked a century ago, when zoo promoters said they were differentiating themselves from menageries by moving toward permanence, education, and science. "Zoos have been transformed, almost overnight, from prisons into bioparks, becoming lush greenswards to enhance the well-being of animals and people," according to Zoo Atlanta's Terry Maple. Recent talk of rehabilitating "scruffy little animal slums" echoes the language of an earlier generation of zoological reformers. Some zoo directors have tried to mark the late-twentieth-century shift with a similar name change: a hundred years ago, menageries became zoological gardens; now the old zoos have become conservation parks and bioparks. These zoo leaders represent the history of zoos as a progression through three stages—menagerie, zoological garden, and conservation park—with an upward-pointing arrow to illustrate inevitable progress in the practice of keeping and displaying animals.[2]

Critics of zoos find this story of progress self-serving and disingenuous, and zoos may have more critics today than ever before in their history. While many aspects of keeping and displaying animals have changed, entertainment generally remains a higher priority than species preservation. ("Animals are fun—especially live ones," according to an advertisement for the Denver Zoo.) Many of the goals to which "post-revolutionary" zoos aspire remain unachieved. In the same period of time that zoos have struggled to reinvent themselves, a group of philosophers and ethicists—every bit as idealistic as the zoo van-guard—has honed arguments against the right of humans even to keep animals in captivity.

Adopting a new name—turning zoos into conservation parks—is a strategy to distance today's displays from the past. By emphasizing discontinuities in philosophy and exhibitry today's zoo promoters dis-miss and discredit historical practices. They contrast their parks to a stereotypical zoo of the past—a "dankly cheerless place of cages, smells, litter, and lost children." This stereotype, however, is a conve-nient generalization. Zoos have always strived to improve animal care and exhibitry. And they have always interpreted nature in complicated and often contradictory ways. Their diverse audiences no doubt have taken home different messages from the experience. The last century of human population growth, wildlife and habitat destruction, and new understanding of ecology and animal behavior has radically altered the way Americans think about nature. Painting the past as uniformly bad makes impossible a nuanced understanding of how zoos have conveyed changes in ideas about the natural world.[3]

Furthermore, the dank-and-smelly zoo belongs to a particular histor-ical moment when genuinely run-down parks coincided with a shift in public attitudes toward animals. The low point against which zoos today measure their progress occurred between the 1950s and the 1970s, as cities entered fiscal crisis and zoo facilities deteriorated from years of neglect. In the first four decades of the twentieth century no-body complained about zoos; even with their animals behind bars, they were civic gems. But in the 1960s the environmental movement began to change public attitudes about wildlife and expectations for how zoo animals ought to be displayed and cared for. By the 1970s an activist animal rights movement nurtured public uneasiness about keeping wild animals in captivity. The ecological idea of biodiversity, and the importance of preserving biodiversity for human welfare, en-tered the popular lexicon. An American population that was increas-ingly urban, watching television shows and films that depicted ani-mals in their natural habitats, looked at zoos and found them lacking. Both the real physical decline of zoos and altered public expectations for animal exhibits pushed zoos to change their exhibits and animal keeping practices.[4]

Pressure to transform zoos into institutions for species preservation and conservation research came from within the zoo community as well as from outside. The New York Zoological Society had for decades supported field research and conservation education. By 1970 a new generation of zoo keepers, involved in the environmental movement and educated in college zoology departments and wildlife management programs, began to replace a zoo keeper work force that had historically been made up of uneducated, unskilled, or semiskilled laborers whose experience in animal management was practical and hands on. Women joined the zookeeping profession in increasing numbers. The better-educated keepers came to work at zoos with experience in research. At all levels the staff of zoos began to be professionalized. Other institutional changes followed: zoos acquired reference libraries for their animal management staff, and public relations and development offices. To offset the lack of municipal funds, zoos began charging admission fees and forming booster groups—zoological societies like the Friends of the National Zoo—to help with fundraising and, increasingly, managing zoos. In 1971 the American Association of Zoological Parks and Aquariums separated from its former association with parks and recreation groups. The independent organization developed an accreditation program and a code of professional ethics.[5]

Despite the talk of revolution, change at zoos has taken place gradually rather than lurching through stages. The evolution of the "new zoo" has steadily gained speed since the end of World War II, with increasing public recognition of the depletion and extinction of wildlife worldwide. At first the changes at zoos were more practical than conservation minded: in the second half of the twentieth century it became much more difficult and expensive for zoos to obtain animals. This prompted greater attention to animal health and well-being and to the systematic breeding of animals in captivity, and an educational focus on conservation. The changes at zoos have served both the ideology of environmentalism and the day-to-day needs of zoos to maintain their collections.

Local and Global Animal Collections

More than anything, what forced innovation at American zoos was the decline of wildlife populations. Long before the widely touted changes of the past twenty or thirty years, zoos began to face challenges in the simple necessity of acquiring animals. World War II marked the end of the era of large-scale animal collecting expeditions. The war disrupted trade routes, and newly independent countries restricted the export of their wildlife. Furthermore, zoos had always been subject to the regulations of the U.S. Department of Agriculture, which required inspection and quarantine of imported animal shipments in order to prevent the spread of contagious animal diseases. In the 1930s and 1940s new species were restricted from importation, and after World War II such regulations increasingly curtailed the wild animal trade. Many animal dealers went out of business. By the 1960s zoos often sent their own staff abroad to make purchases and escort animals home. Tigers, rhinoceroses, and other species had become too rare and too valuable to trust to the vagaries of shipping and handling.[6]

Still the attitude persisted at zoos that acquiring animals "was a business proposition. Animals were there. You could get them, and you didn't worry about where they were coming from." Zoo directors operated on the premise that, "We'll have animals forever. Replacement's easy." It was also cheaper to replace animals as needed rather than to keep unwanted "surplus" stock in the zoo. It took a ground swell of public and scientific concern for the future of wildlife, and the passage of federal legislation, to make zoo managers realize that, if animals were endangered, so were zoos.[7]

Between 1966 and 1973 the U.S. Congress passed three endangered species acts, under pressure from environmentalists, wilderness advocates, wildlife protection groups, and humane societies, which recognized the imminent extinction of both domestic and foreign species. In 1972 the Marine Mammal Protection Act placed severe restrictions on commercial trade in animals like sea lions, that were staples of zoo collections. A year later the United States signed the Convention on International Trade in Endangered Species of Wild Fauna and Flora (CITES). In addition, the Animal Welfare Act, first legislated in 1966, mandated standards of care for zoo animals as well as yearly inspections.[8]

Municipal zoos, which were used to dealing with only local politics, were thrust into federal bureaucracies by this new legislation. Suddenly they were being inspected by the Department of Agriculture, applying for permits from the Department of Commerce, and submitting reports to the Department of the Interior. Meeting standards of care and adhering to new federal regulations regarding endangered species was expensive, and zoos were being required to upgrade facilities and step up research and conservation efforts at a time when municipal governments could offer little support. In fact, political support for cultural institutions in general was shrinking. Tensions ran high between city park administrators who saw zoos mainly as recreational centers and zoo staff required to meet new federal regulations. For most zoos, channeling resources to conservation efforts remained a dream for the future. As one zoo curator has described it, "The thought of spending municipal tax dollars to fund research and conservation programs for wildlife indigenous to Africa, Asia, or South America was ludicrous."[9]

Despite increased regulation, there were still loopholes through which endangered animals could find their way into American zoos. A zoo might be allowed to import golden lion tamarins from Colombia, for example, but not from their native Brazil. So if the tamarins had been transported from Brazil to Colombia, and then to the United States, importing them was legal. But such animal-laundering practices clearly could not be sustained, nor did they diminish the fact that wild populations were endangered. Furthermore, in the 1960s public opinion turned against the wild animal trade, and articles in the popular press drew attention to the astonishing loss of life it perpetuated. Frank Buck's stories in the 1930s had matter-of-factly detailed animal deaths on the journey from jungle to zoo, and they provoked no complaint. The public mood had changed by the time of a 1968 article in *Life* magazine. Illustrated with photographs of animals that had suffocated in transport and others crammed into small cages, it declared, "The enormous and profitable traffic in wildlife—for food, sport, skins, zoos, scientific research and even pets—decimates whole species and threatens to wipe out those rare specimens from which man derives such benefit and delight."[10]

If zoos did not start systematically breeding their own animals, they would not have animals to display. Zoos also risked losing their audience if they did not convince the public that they were part of the effort to save species rather than contributing to extinctions. The wildlife legislation of the late 1960s and early 1970s prompted an inventory of animals owned by zoos, and the application of the latest knowledge in genetics and reproductive science to launch a nationally organized system for captive breeding.[11]

Some species had bred in captivity long before it became a matter of conservation. For over a century births at the zoo gave zoo personnel the opportunity to learn about gestation periods and animal development. Frank Thompson, for example, published his observations of black bear cubs born in the Cincinnati Zoo in 1879, beginning three days after their birth. He had walked into the enclosure with the bears' keeper, who "quickly began feeding her with bits of bread, which he sliced from a loaf held in his hand. By holding the bread just over her head, he finally tempted her to sit up on her haunches, when I obtained a clear view of the two young ones, lying asleep just back of her front paws." Thompson made note of their color, the number of days until their eyes opened, when they emerged from the den, and how they "soon became expert climbers." According to the magazine *Forest and Stream*, Thompson's article was "the first account that we have ever read of the babyhood of bears."[12]

Lions also reproduced readily in captivity; a lioness at the Memphis Zoo, for example, raised sixteen cubs in the early 1920s, and the zoo also bred zebras and hippopotamuses. Some zoos had success with particular species. The Milwaukee Zoo was known in the 1920s for breeding and raising polar bears. Nine cubs were born to one female, named Sultana, between 1919 and 1929. Giving credit where it was due, the zoo's director in 1932, Edmund Heller, wrote that Sultana's "affection and solicitude for her cubs is intense and she deserves the major portion of the credit for the wonderful achievement of raising polars in captivity." Others also observed that, although animals often bred in the zoo, "it is somewhat of a proposition to raise the young successfully."[13]

When new mothers at the zoo were disinclined to nurse, or seemed likely to hurt their newborns, keepers took the babies away and attempted to hand-rear them, or rather, they took them home for their wives to raise. Zookeeping was not a profession open to women in the United States until the late 1960s, but many a lion cub was bottle-fed by a keeper's wife. At the Houston Zoo, a "mother chimpanzee failed to take care of her baby and so Mrs. Nagel has taken the animal to her home where she will give it as much care and attention as a child would receive." Helen Martini, who raised dozens of animals in her apartment near the Bronx Zoo, wrote a book about her "zoo family," and in 1959, the National Zoo honored five wives and one mother of zoo employees with a luncheon for their work in raising zoo babies.[14]

If for the most part serendipitous, "births and hatchings" at zoos were nonetheless important. Only animals in good health would breed in the zoo, and so zoo births were taken as evidence that the animals were receiving good care. Baby animals also drew crowds to the zoo—they provided a new exhibit practically for free. Zoo nurseries and children's zoos with baby animals became popular attractions in the 1950s. Finally, animals born in the zoo became a source of revenue when sold to other zoos, dealers, or private collectors.

It took time to convert serendipitous births into managed breeding programs. Although individual zoos kept records on animal breeding and the pedigree of animals born in the zoo, this information was not exchanged when animals were sold or traded. Zoos did not coordinate their records, and as they stepped up their efforts to increase zoo births, this could lead to inbreeding. European zoos were more attentive to record keeping, and to studbook keeping for rare animals, and the problems created when zoos lacked records were first observed there. So little was known about managing exotic animals that into the 1970s American zookeepers widely believed that such animals could not become inbred and suffer consequent health problems. Americans became more aware of problems of inbreeding through a study published in 1979 by researchers at the National Zoo. They reported that, in sixteen species of hoofed animals, the offspring of parents that were related were significantly less likely to survive.[15]

Captive breeding to maintain zoo populations and to build up populations of endangered species would require zoos to improve their record keeping and to share information that zoo directors had traditionally kept to themselves. To compile a comprehensive tally of zoo inhabitants and their vital statistics, the American Association of Zoological Parks and Aquariums (AAZPA) (now called the American Zoo and Aquarium Association), among other supporters, endorsed the creation of an International Species Inventory System (ISIS) in 1973. Operated out of the Minnesota Zoological Garden, ISIS began with a census of mammals held in North American zoos. With detailed information about species held, the ISIS managers could begin to figure out whether self-sustaining populations of endangered species existed in zoos. The creation of ISIS signaled a shift toward long-term genetic and demographic planning for the animals in American zoos. Since its inception, ISIS has grown to track hundreds of thousands of specimens around the world, and has developed specialized databases for medical and genetic information.[16]

In 1981, the AAZPA started a second program that built on the ISIS information—a series of Species Survival Plans (SSPs) aimed at saving chosen species from extinction through captive breeding, habitat preservation, and supporting research. These survival plans are programs for the long-term survival of species. Zoos participate voluntarily to exchange animals for breeding in the interest of maintaining genetic diversity in the population of captive animals. By 2001 there were Species Survival Plans for 139 species, including the giant panda, the okapi, the Bali mynah, the Partula snail, and the Puerto Rican crested toad.[17]

Zoo science now includes genetic analysis, to distinguish closely related subspecies, for example, and the full menu of reproductive technology—embryo transplants, in vitro fertilization, and banks of frozen sperm, embryos, and animal tissue. Cloning has been suggested as a possible means of reproducing near-extinct animals in the future. Accomplishing this work has required zoos to look beyond the borders of their parks. Since many animals will not breed in confined conditions, before the gaze of the public, a few zoos have established breeding and research stations on thousands of acres of land, far removed from their urban parks. Both on their own and through collaborative programs, American zoos also support wildlife and habitat conservation efforts around the world. Zoo professionals unify their wide-ranging laboratory and field studies of animals under the umbrella of "zoo biology"—a field that "embraces everything in the zoo which is biologically relevant . . . from zoology to human psychology and from ecology to pathology."[18]

Researchers at zoos see the animals as belonging to worldwide populations rather than as the pets of a particular city; individual animals have become representatives of gene pools. Administrators, keepers, and researchers who are better educated than their predecessors and who take the conservation mission of zoos seriously have accompanied this transformation. In thirty years, captive breeding has grown from an effort to keep a stock of zoo animals for display to a plan for maintaining biodiversity in both zoos and natural habitats.

The turn to captive breeding has been both pragmatic, in helping to guarantee the future of zoos, and altruistic, in serving the cause of species preservation. It has also provoked a host of questions. With shrinking habitats, should zoos breed animal populations that have no home to return to? If they are released from the zoo, can captive bred and raised animals survive and reproduce in the wild? If a large population of animals is needed to maintain genetic diversity, but zoos can keep only a small number of them, what happens to the "surplus"? How much of animal behavior is learned socially, and thus lost if a generation of animals is stored as frozen embryos? Furthermore, despite zoo researchers' best efforts, many species have not bred in captivity, and even the most sophisticated technology cannot succeed given the limited knowledge available about the reproductive biology of most animals. The focus on breeding animals and managing populations has opened avenues of scientific research for zoo staff, but it has also made them vulnerable to criticism.

Animal Health and Animal Welfare

Zoo keepers and directors have always been concerned with maintaining the health of their animals. At the turn of the twenty-first century they have begun to refer to efforts at improving the health of zoo residents as "animal welfare science." In the community of zoo professionals this refers to systematizing care and monitoring physiological indicators of stress in order to put on a scientific basis what so-called gifted keepers have always done. It also represents a shift toward focusing on the daily health of individual animals rather than the long-term genetic viability of populations. But before examining the new animal welfare science, it is worth taking a brief look at some efforts to improve animal care a century ago.

The Philadelphia Zoo was among the first American zoos to encourage new research directly related to controlling animal disease—specifically, to eliminate tuberculosis infection in the zoo. Beginning in 1901 the zoo conducted necropsies (animal autopsies) of all animals that died there. The pathology laboratory eventually became known as the Penrose Research Laboratory, after Charles B. Penrose, a professor at the University of Pennsylvania medical school who started the work with Ellen P. Corson-White. The necropsies showed that tuberculosis was a chronic health problem, particularly among the primates in the zoo. Joined by Herbert Fox in 1906, the researchers developed a skin test for tuberculosis that became widely used on animals, and set up a system of quarantining and testing all the animals in the monkey house. They also installed glass fronts on the cages to prevent transmission of the disease between visiting members of the public and the animals. By 1909 they claimed that the monkey house had been free from tuberculosis for two years, although the disease remained a recurring problem.[19]

In addition to tuberculosis, the necropsy records revealed that malnutrition was a frequent cause of death for zoo animals. Little was known about what animals ate in the wild. Collectors generally kept baby animals alive on a gruel of goat's milk and rice until the animals were sold. Once at the zoo, they often were fed according to traditions inherited from the keepers of circus animals. By custom, lions were given raw beef or horse meat six days a week, and fasted on the seventh. But not all zoo animals thrived. "Cage paralysis," a weakness of the hind limbs similar to rickets, often developed in monkeys and big cats. Researchers at the Philadelphia Zoo in the first decades of the twentieth century worked to understand and improve the diets of zoo animals. They eventually developed "zoo cake," a mix of grains and oils that was fed with supplements to most animals in the zoo. Other zoos imitated and refined the idea. The Bronx Zoo, for example, devised "a kind of forage biscuit, made of a mixture of bran, oats, corn, molasses, bone-meal, salt and other minerals," to feed to certain antelopes.[20]

The Philadelphia Zoo was exceptional in the United States in having professional staff dedicated to improving animal health. More typically, the keepers were in charge of day-to-day animal care. Some keepers came to this work with experience caring for circus or farm animals. Others were parks department employees hired as laborers. In either case, through most of the twentieth century many keepers stayed in their jobs for thirty or forty years or more. They knew their animals, and zoo directors often deferred to them in matters of animal health. When William M. Mann summed up the state of research at the National Zoo in 1950, he was describing the work of a staff of attentive animal keepers:

> Scientific research is not set up as a separate activity in the National Zoological Park but is an important part of the operation. The proper care of hundreds of different kinds of animals, some of which have not previously been kept alive, calls for constant observation and study to determine for each one its natural living conditions, likes and dislikes. Usually the most important step is to try foods that will be acceptable substitutes for those that the animals would normally obtain in the wild. Other conditions, such as humidity, temperature, type of bedding, types of perch, indeed everything affecting the animal in captivity, require constant study to make certain that a suitable environment is maintained.[21]

Although many keepers kept journals in which they described their daily work with the animals, research of this sort generally was not written up and published. Zoos competed fiercely with each other to exhibit the greatest variety of species, and to set records for their longevity. They considered successful diets and other keeping practices to be proprietary knowledge.[22]

Attending to sick zoo animals was just as limited as human medicine before the discovery of antibiotics, with the added complication that it could be difficult to examine the patient. Keepers induced big cats to take castor oil, for example, by pouring it from above onto their heads and fore paws, so the animals would lick it off. Pills could be ground up and mixed into food. When more active intervention was called for—surgery, for example—a zoo might hire either a veterinarian or a physician for individual cases.[23]

Few zoos had veterinarians on staff until the 1960s. Instead, they often developed a long-term relationship with a local veterinarian or physician who would treat animals when the zoo director deemed it necessary. Bringing in an outsider could provoke considerable conflict. There was a healthy and mutual disrespect between veterinarians, who were trained to treat farm animals, horses, and pets, and who did not know the physiology of exotic animals in detail, and keepers, who rarely had more than a high-school education, but were familiar with the animals' day-to-day habits. As Belle Benchley, director of the San Diego Zoo, put it in 1940, "The keepers felt that if they permitted him [the veterinarian] to get his hands on their darlings, he would kill them with his queer book-learning. . . . then I found that an attitude of contempt for the seeming illiteracy of some of our best men on the part of one veterinarian was causing him to treat their suggestions and reports with open scorn." This rift persisted for decades.[24]

Calling in a veterinarian or physician to perform surgery was something of a last resort for other reasons as well. Unless an animal was tame, attempting to treat it could put the lives of both surgeon and animal in danger. The stress and struggle involved in restraining an animal could harm it more than disease. "If the animal happens to be a buffalo bull," wrote the Bronx Zoo's veterinarian Reid Blair, "it is driven into a small enclosure, lassoed, thrown down, and tied by its three well feet, while five or six men sit on the struggling beast." Bones could easily be broken during this rough process. Other animals were also difficult to restrain. With typical understatement, Blair commented, "Preparing a twelve-foot alligator for an operation is a risky task."[25]

The development of antibiotics in the 1940s changed animal health care just as it changed human health care. Getting the drugs into the animals was still a challenge, however, and a new instrument called the Cap-Chur gun helped solve the problem in two ways. It could be used to administer, from a distance, either a tranquilizing drug that made dangerous animals tractable or the antibiotics themselves. Basically an air gun that could fire a syringe, the Cap-Chur gun was first developed by game managers in order to tranquilize, capture, and relocate deer.[26] Veterinarians at the National Zoological Park were the first to try it on zoo animals, and in 1959 Theodore H. Reed, the director of the zoo, introduced the gun to the wider zoo community at a meeting of the International Union of Directors of Zoological Gardens. Over the next decades researchers at the National Zoo and elsewhere tested and refined new tranquilizers and dart guns.[27]

Advances in veterinary medicine, including better drugs and dewormers, have continued to be applied to the health care of zoo animals. Recently, diagnostic imaging techniques such as ultrasound, and MRI and CT scanning, have become valuable tools for zoo veterinarians. Zoo hospitals are now equipped to perform endoscopy and fiberoptic surgery. Just as for humans, such technologies improve the diagnosis and treatment of disease. Better health care for animals also has aided efforts to take down the bars from zoo displays and give animals larger and more naturalistic living quarters—antibiotics and veterinary expertise help keep animals healthy in these less than sterile, and potentially hazardous, surroundings.[28]

More Natural Settings

The most visible transformation at zoos in the last thirty years is in the landscape. Like changes in animal keeping and breeding, the way zoo exhibits look has been influenced by the science of ecology, the increasing urgency of conservation, and concern for animal welfare. Landscape immersion—a term coined by landscape architects to describe exhibits that attempt to envelop zoo visitors in animals' environments rather than setting animals down in parks with trim lawns and potted plants—became the industry standard by the 1980s. The concept was developed during the wholesale renovation of the Woodland Park Zoo in Seattle in the late 1970s. The zoo's master plan stated its goal succinctly: "Wild animals live in a dynamic ecological relationship, and, although the zoo is only a substitute, it should attempt to reflect this complex order."[29]

The idea that zoos should display ecology rather than taxonomy had been gaining momentum for decades. On a smaller scale, some zoos had attempted to show ecological relationships in their displays as early as the 1940s. The Bronx Zoo's African Plains exhibit, which opened in 1941, was perhaps the first. Earlier in the century, William Hornaday had objected to exhibits with moats patterned after those of Carl Hagenbeck, and none were built in the Bronx Zoo. After Hornaday's retirement, however, the African Plains was planned, and it took this style of display beyond the theater of showing predators and prey near each other. Hagenbeck's displays often combined species from different continents. In contrast, wrote Fairfield Osborn, president of the New York Zoological Society, in "Africa in the Bronx" and future exhibits "the animal collections, to the greatest degree possible, shall be shown grouped as they are in nature." The educational mission of such exhibits in Osborn's view was to encourage audiences to think of humans as part of natural systems. He wrote: "if man is to fulfill his potential destiny, he must give thought to his relationship to nature— to his dependence upon all the forms of life that surround him."[30]

The idea of conservation education at zoos also has a longer history than is often acknowledged. This idea was foremost among the goals of the Arizona-Sonora Desert Museum as it was planned in the early 1950s. Animal exhibits were one component of the museum, which was begun with the goal of educating the public about the plant life and scenic value of the desert, as well as its wildlife. Although the museum's focus was regional, and it was not a traditional zoo, its displays were designed to draw attention to connections between animals and their environments, and zoo directors looked to it as a model.[31]

In the meantime, in Milwaukee, Wisconsin, an effort was under way to build a new zoo that would "exhibit wild animals to the greatest degree possible as they occur in nature" on a 180-acre site outside the city. And in San Diego the idea for the Wild Animal Park was first discussed in the early 1960s; the park opened in 1972. By the late 1960s some zoo directors were using the word "revolution" to describe the scope of changes they planned for their institutions. In 1968 William Conway, then director of the New York Zoological Society, published an article that has set the agenda for exhibit design and educational programming up to today. Titled "How to Exhibit a Bullfrog: A Bed-Time Story for Zoo Men," the article elaborated on an imaginary exhibit that put the mundane bullfrog in a natural setting with other wildlife, incorporated video cameras and binoculars to give visitors close-up views of animals, and used graphic design, sound, print materials, and film to educate visitors about natural history, evolution, and conservation.[32]

The lesson for the zoo community was that displays ought to be about more than the animals themselves in order to shape visitors' political behavior: "it is the Bronx bus driver and the corner pharmacist whose votes will determine the fate of the Adirondack wilderness, the Everglades, of Yellowstone Park," Conway wrote. "You must give your visitors a new intellectual reference point, meaningful and aesthetically compelling; a view of another sensory and social world; a feeling of personal interest in diminishing wild creatures and collective responsibility for their future which is so closely linked to that of man. Zoos must be natural history and conservation centres for the future." The Bronx Zoo was already moving in this direction with its World of Birds and World of Darkness exhibits, which displayed animals in a way that emphasized habitats and incorporated educational tools such as recordings of animal sounds and automated slide and film presentations.[33]

Twenty years later Conway's vision was so widely accepted that much of it was incorporated into a textbook. The transformation in exhibit design accelerated in the 1980s and 1990s when a good economy made funds available for new exhibits and as a specialized community of architects and landscape designers gained expertise in carrying out the work. Indeed, landscape architects sometimes claim credit for the sea change in zoo displays. New exhibits shaped the human experience of nature in the zoo, manipulating sight lines, for example, so that visitors looked up at animals rather than down on them in reverse of the traditional dominance hierarchy. Plantings were arranged to educate and to enhance the visitor experience, for example, by demonstrating "the function of the leopard's distinctive coloration."[34]

Immersion exhibits were assumed to improve the quality of life for animals. But zoo curators recognized that " 'naturalistic' from the visitor's point of view . . . is not necessarily naturalistic from the animal's point of view." In the 1990s they began to pay more attention to "enriching" the lives of animals in these and in more traditional displays. Programs for environmental enrichment, also referred to as behavioral enrichment, are aimed at improving both the physiological and psychological well-being of captive animals. They stimulate animals to exercise and elicit behaviors from animals like those they show in the wild; enrichment is thus a way to train animals that will be reintroduced to wild habitats in survival skills. The simplest enrichment techniques include scattering food around enclosures so that the animals forage and explore rather than eat from a bowl, and providing toys that stimulate activity and relieve stress. Interaction with keepers can also be a form of enrichment; training an animal to lift a foot or turn its head for a keeper to examine, for example, both engages the animal and makes veterinary care easier. Ideally, animals with appropriate enrichment behave much as they would in nature and do not pace the perimeters of their enclosures or display other signs of stress. All this adds up to what zoo curators call "animal welfare science"— a balancing of concern for maintaining populations with the well-being of individual animals, and an effort to measure and quantify the relationships between day-to-day nutrition and care and animal health and behavior. Animal welfare science has been institutionalized in the American Zoo and Aquarium Association, which held the first meeting of its Animal Welfare Committee in 2000. The committee's charge is both humanitarian—to ensure that animal welfare is a priority at its member zoos—and public relations—an effort to assure the public that zoos are good places for animals.[35]

By the end of the twentieth century, zoo managers had rearranged the order of the institutional goals with which they were founded. Whereas entertainment once led the list of zoo goals, today's mission is conservation, education, science, and recreation. The American Zoo and Aquarium Association "has adopted conservation of the world's wildlife and their habitats as its highest priority." Exhibit designer John Coe goes a step further, writing, "The ultimate goal is to increase public awareness and appreciation of the importance of habitat and its protection to wildlife conservation and to present zoo animals in such a way that their reason for being and rights to existence are intuitively self-evident to viewers." But at the same time that the zoo community has stepped up efforts at species and habitat preservation, and increased attention to individual animal welfare, a portion of the audience for zoos has become ever more uneasy with the practice of keeping wild animals in captivity. Zoo curators admit that "we more often find ourselves on the defensive than on the offensive with respect to issues involving animal welfare." Concern for animal welfare and animal rights has a history that has often been at odds with zoos.[36]

Animal Welfare and Animal Rights

A portion of zoo observers, although small during most of the last century, has always questioned the ethics of keeping wild animals captive. Such anxiety has often been passive, in effect condoning the existence of zoos. Leonard Woolf expressed the sentiment well when he wrote, "I am ambivalent about zoos: I have an uneasy feeling that one should not keep animals in cages, but I never get tired of watching animals anywhere."[37]

American zoological parks came into existence in the late nineteenth century at the same time as a reform movement for animal welfare. Like other urban humanitarians, animal welfarists were concerned with eliminating suffering. They formed animal protection societies, including the Society for the Prevention of Cruelty to Animals in 1866. Guided in thought and practice by the English humanitarian movement, American animal protectionists directed their campaigns mainly at improved treatment of domesticated animals such as dogs and working horses, and at protesting vivisection in scientific research. Occasionally they expressed concern for wild animals as well—they rallied against sport hunting, trapping animals for furs, and killing birds for the plume trade.[38]

A few humanitarians spoke out against zoos, although they formed no organized opposition. Their criticisms centered on two issues—the welfare of the animals and their lack of freedom in the zoo. In terms of welfare, their concerns—like those for other animals—were that zoo animals were healthy and well cared for, and not subjected to cruelty or pain. The practice of keeping wild animals confined also provoked occasional objection. In the context of zoos this idea was not well articulated beyond the feeling that "wild animals ought to roam free," but it expressed a notion about authenticity in the natural world—that wild animals deserved to live an authentic existence in nature. Such empathy led some writers early in the twentieth century to propose that animals are entitled to moral status, even rights such as liberty, although they were not writing about animals in zoos. These two concerns about animal welfare and authenticity have carried through a century of critiques of zoos.[39]

Humanitarians protested cruelty in training animals for circuses more often than they opposed zoos. Hundreds of thousands of people sent in their names to the Jack London Club in a show of support for boycotting animal entertainments. The club was promoted in the pages of *Our Dumb Animals*, the newsletter of the Massachusetts Society for the Prevention of Cruelty to Animals. But the same publication scarcely mentioned zoos. The local Franklin Park Zoo, which opened in Boston in 1914, had been a subject of public discussion as it was planned, but it was not a concern of *Our Dumb Animals*. And, although opposed to zoos, the editor praised the Boston *Post*'s campaign to raise money to purchase three performing elephants for the zoo based on the reasoning that life in the zoo would be less cruel than life on the stage—the zoo would "release these noble animals from the hardships of travel and the continued performances at exhibitions." The article continued, "We do not believe in the captivity of wild animals, but since these elephants are here, and there is no opportunity to send them back to their native haunts (and doubtless that would be unwise now even if possible), we rejoice that they are to have such comforts as a well-managed zoological garden can give them." Lacking other alternatives, keeping animals in a zoo was preferable to the indignity and, potentially, abuse of life on the vaudeville stage.[40]

There is no question that zoo critics were in the minority at a time when a new zoo was the pride of every forward-thinking city. But to judge by the frequency with which zoo boosters expounded on the well-being of zoo animals, they must have felt vulnerable to reproach. New displays, like the naturalistic bear exhibit completed at the Denver Zoo in 1918, were applauded as humane among their other merits. Zoo promoters tended to rely on two arguments supporting animal welfare in the zoo. First, they said that the laws of nature—meaning the daily struggle for existence and avoidance of predators—were far more cruel than conditions at the zoo. A contributor to *Scientific Ameri-*

can made a typical argument in 1915: "Remember that everything in the jungle is in danger, 'everywhere stalks the grim specter.' In the zoo not an inmate of any kind has a single enemy to fear." Evidence that animals lived healthier and longer lives in the zoo than in the wild supported this line of thinking. Ironically, a British game hunter, F. C. Selous, was often quoted as an authority on the subject. Selous purportedly could distinguish the prepared skin of a "menagerie lion" from that of a wild one because it was healthier and had a longer and glossier coat.[41]

The second argument supporting animal welfare in the zoo responded to visitors' worries about the loss of freedom of zoo animals. While zoo supporters agreed that animals were capable of experiencing physical pain and suffering, they did not believe that animals possessed the mental capacity to suffer anguish from the mere fact of captivity: "An animal does not have a man's hopeless feeling in confinement, he knows he is restrained for the time being, but he does not feel the shame and disgrace, the shattered hopes and ambitions of the human prisoner." They did not consider animal and human minds equivalent, and they drew a firm line between humans and animals on the issue of the moral status of animals.[42]

In the 1950s zoo professionals added scientific backing to their conviction that keeping animals in captivity did not cause them undue stress. The Swiss zoo director and animal behaviorist Heini Hediger had published the book *Wild Animals in Captivity* in 1942, and in 1950 it was translated into English. Discussing concepts such as flight distance, geographic range, ecological niche, territory, predator-prey relationships, and biological and social rank in animals, Hediger argued that for animals in captivity, it was the quality of their living space rather than the quantity of it that was important. Field research had established that—contrary to the sentimental idea that wild animals roamed free—many animals established territories. They lived within confined areas that they defined and marked for themselves. Extending the idea of territory to zoo animals, a fenced paddock could serve just as well as a territory in the wild, so long as it provided enough space for animals to retreat from humans or other animals who might be perceived as predators. What wild animals really craved was security, and this the zoo provided. In the glib assessment of *Time* magazine, which reviewed Hediger's book, "If properly housed, fed and entertained, [animals in zoos] often lead happier, fuller lives than the humans who come to watch them."[43]

Through the 1950s and early 1960s articles like Fairfield Osborn's "The Good Life in the Zoo," and "Don't Pity the Animals in the Zoo," an interview with William M. Mann, repeated these familiar arguments: "all animals in the wild are engaged in a struggle for survival, and many live in constant fear, or suffer from starvation, injury or disease," wrote Osborn. Ironically, Hediger's ideas about animal behavior were used to support a particularly sterile mode of exhibitry, intended to protect animals from infections. Zoo managers today refer to such displays as the "bathroom period." Lined with tiles so that they could easily be hosed down and disinfected, such minimalist cages were defended by saying that the animals had plenty to eat and nothing to fear.[44]

Although practical from the point of view of cleanliness and durability, such "naked cages" also provoked protest. In many cases these are the exhibits in the "bad old zoos" against which habitat displays are pitched. Animals paced neurotically or mutilated themselves out of frustration and boredom. Visitors reacted to the starkness of the cages, lamenting the animals' loss of freedom. Zoo directors too were distressed by the behavioral abnormalities put on display, and started agitating for more complex, less uniform, exhibits in order to relieve the stress of captivity. By 1968, London Zoo curator Desmond Morris was calling for a zoo revolution to make exhibits more naturalistic and complex for the mental well-being of the animals.[45]

In the meantime, a different revolution was brewing: a new, activist animal rights movement. Humane societies had remained active throughout the century, devoting their efforts to shelters for stray pets and passing leash laws, and in the 1950s and 1960s protesting against stray dogs and stolen pets being sold to university laboratories or pharmaceutical companies. New groups were formed in these decades, however, and they began to use more aggressive tactics than their predecessors. The movement that emerged in the 1970s went beyond the traditional welfarist concern of protecting animals against cruelty by demanding rights for individual animals. Although the many rights groups have never been uniform in their aims, they find common ground in philosopher Peter Singer's book *Animal Liberation*, published in a paperback edition in 1977. In fact, so many animal rights activists own the book that its sales have been used to track the growth of the movement. Among other things, Singer argued that animals are worthy of moral consideration, and that to deny their suffering amounts to "speciesism," an offense much the same as racism or sexism. Animal rights advocates reiterate this view. "Just as there is no master sex and no master race," writes Tom Regan, "so, (animal rights advocates maintain) there is no master species."[46]

In the early 1990s a group of primatologists, anthropologists, psychologists, ethologists, and ethicists collectively supported a more limited extension of rights to the mammals most closely related to humans—and among the most popular at zoos—the great apes. Led by Singer and Paola Cavalieri, they subscribed to "A Declaration on Great Apes," which demanded that gorillas, chimpanzees, and orangutans be included in the "community of equals" with humans. Specifically, they advocated extending the right to life, protection of individual liberty, and prohibition of torture to the great apes.[47]

Although in the United States zoos have not been the most visible targets of animal rights activists, the rights philosophy in relation to zoos has been clearly articulated. Philosopher Dale Jamieson, perhaps the most outspoken on the issue, argues that a moral presumption against keeping animals in captivity outweighs any benefit that might accrue from education, entertainment, science, or species preservation. Jamieson and others also contest the ability of zoos to preserve species through captive breeding programs. At its core, the rights philosophy opposes keeping animals in captivity under any conditions, and argues for abolishing zoos. It is the late-twentieth-century elaboration of the earlier idea that "wild animals ought to roam free," a statement about inherent value in the authentic state of wildness. Other zoo critics examine authenticity from a different angle, worrying about the effect on visitors of looking at animals in artificial settings. "The spectator does not see a zebra in the zoo—a zebra is something that exists on an African plain, not in an urban North American animal collection," writes Randy Malamud. In addition to violating the rights of animals, Malamud argues that "zoos represent a cultural danger, a deadening of our sensibilities." From either perspective, the requirement of authenticity for animal rights leaves no room for negotiation about zoos. Nonetheless, the rights critique has helped prompt zoo professionals to reflect on the conflicting values at issue in wildlife management, and to seek some consensus on animal welfare.[48]

The animal rights movement represents one end of a broad spectrum of changes in American attitudes toward animals in recent decades. The human-animal relationship, in particular its potential for the exploitation of animals, has become problematic in ways that it never was before. The increase in pet keeping in industrialized countries and changes in attitudes toward pets provide one example. The word "pet" itself is contested; many dog and cat owners prefer the term "companion animal," and a few American cities have passed ordinances making humans the "guardians" of their animals. This redefinition of long-standing hierarchies deserves more study than it has been given. No doubt it influences current public attitudes toward zoos, and will shape the future of zoos.[49]

Wild Animals in a Crowded World

In the century or so since American zoos were founded, the natural world has been transformed. Animals in the zoo no longer stand in for large populations of their kind still living somewhere in nature, innocent of Western civilization, awaiting discovery by an intrepid collector. Human populations have encroached on most corners of the world, displacing animals, and "wild" places have been fenced in to keep humans out. Many animals in the zoo that were considered rare in the past because few people in North American cities had seen them are now genuinely rare—they are among the last representatives of their species. In response, zoos have changed how they assemble their collections and keep their animals, shifted and strengthened their educational mission, and undertaken new research.

Yet zoological parks adhere to their original mission: recreation, education, conservation, and scientific research. The sweep of these goals helps account for the lasting popularity of zoos—zoos have something for everyone. But these goals are also fraught with conflict. Exhibits of animals can't help but objectify the natural world, presenting wildlife to be observed and enjoyed, even as habitat settings and educational signage soften and shape that message by emphasizing ecological relationships and species extinction. Science raises as many questions as it answers. And zoos convey conflicting messages in their education, marketing, and public relations efforts. At the same time as they seek to educate visitors about authentic animal behavior, they encourage visitors to anthropomorphize through publicity surrounding births at the zoo and in advertisements. An ad for the Bronx Zoo invites visitors to see its "congo line" with a photo of chimpanzees in dance formation.

Conflicts between recreation and education, lowbrow entertainment and highbrow research goals, were publicly and poignantly clear in the 1990s as the New York Zoological Society/Wildlife Conservation Society struggled to rename its parks. Bronx Zoo administrators have been uncomfortable since Hornaday's time with the connotations of the word zoo—confusion and disorder. In 1993 they attempted to change the institution's name to the International Wildlife Conservation Park, "highfalutin words for a highfalutin objective," according to William Conway, then president of the New York Zoological Society. Editors of *The New York Times* were quick to point out that "highfalutin" means "pompous or pretentious." Such posturing was plainly out of place. One commentator added that when New Yorkers "wanna go to the zoo, they'll go to the zoo, y'know what I mean?" The Society eventually relented and put "Bronx Zoo" back into the park's name.[50]

Zoos display the diversity of ways that people appreciate the natural world, and many of their contradictions. Because of this they remain important and enduring institutions. The pleasure of walking outdoors in a park, the possibility of transcendence through meeting the gaze of another species, the opportunity literally to touch nature in petting zoos, the chance to learn to distinguish frog songs or to understand the social structure of a gorilla family—all of these are accessible at the zoo. Zoos can show the charm and deep cultural associations of nature— they present endless possibilities for punning, for "panda-monium," for example, and for small jokes like the National Zoo's long-standing declaration that "lost children will be taken to the lion house." And at the same time they can convey the gravity of environmental destruction, the complexity of the human impact on the natural world, and the urgency of coming to grips with the meaning of these changes.

Zoos also force conflicts about the relationships between humans and wildlife into public debate. Because of their educational mission they can provide alternatives to other opportunities for interacting with wildlife—scripted animal entertainments, private game hunting on Texas ranches, and exotic pet keeping, for example. Zoos keep important discussions about wildlife and habitat conservation visible and give hundreds of millions of people the opportunity, however flawed, to engage with the natural world.

In the brief course of the history of American zoos, the United States changed from a rural agricultural society to an urban industrial one. At the same time, science has developed quantitative and complex explanations for the workings of the natural world. Along the way, human relationships to animals and to nature also have evolved in ways that are only beginning to be grasped. As contrived meeting places for the most urban people and the most exotic animals, zoos have often been a setting for negotiating these relationships. They display the diversity and the complexity of human ideas about animals— their richness, contradiction, and history. In a changed world—ever more populous, developed, and deforested—this may be their most enduring value.

NOTES

Introduction

1. For accounts of Willie B., see, for example, Francis Desiderio, "Raising the Bars: The Transformation of Atlanta's Zoo, 1889–2000," *Atlanta History*, 2000, 43(4):7–51; Terry L. Maple, *Zoo Man: Inside the Zoo Revolution* (Atlanta: Longstreet Press, 1993). For zoo attendance figures, see the web site of the American Zoo and Aquarium Association, www.aza.org.

2. For accounts of animal collections from the ancient world to the present, see, for example, James Fisher, *Zoos of the World: The Story of Animals in Captivity* (Garden City, NY: Natural History Press, for the American Museum of Natural History, 1967); Lynne Iadarola, "Zoos," in *Encyclopedia of Architecture: Design, Engineering, and Construction*, Vol. 5 (New York: John Wiley, 1989), pp. 419–439; Gustav Loisel, *Histoire des Menageries de l'Antiquite a Nos Jours*, 3 vols. (Paris: Octave Doin et Fils, 1912); Stephen St. C. Bostock, *Zoos and Animal Rights: The Ethics of Keeping Animals* (New York: Routledge, 1993).

3. Throughout this study, I use the word "zoo" interchangeably with the more cumbersome terms "zoological park" and "zoological garden." Statistics on the number of zoos in the United States are provided in L.H. Weir, *Parks: A Manual of Municipal and County Parks*, Vol. 2 (New York: A.S. Barnes and Company, 1928), pp. 904–905; and in Vernon N. Kisling, Jr., ed., *Zoo and Aquarium History: Ancient Animal Collections to Zoological Gardens* (Boca Raton, FL: CRC Press, 2000), pp. 375–380. For a general introduction to the history of American zoos, see Helen Lefkowitz Horowitz, "Seeing Ourselves through the Bars," *Landscape*, 1981, 25(2):12–19.

4. For Bronx Zoo attendance figures, see New York Zoological Society, *Eighth Annual Report* 1903:50; New York Zoological Society, *Fourteenth Annual Report*, 1909:108. Toledo figures are in Frank L. Skeldon, "An Educational Mission: Toledo Zoological Gardens Valuable Service and Pleasure," *Parks & Recreation*, 1928, 11(4):285.

5. The tension between popular and scientific understandings of nature was also evident in wildlife films in the early twentieth century. See Gregg Mitman, *Reel Nature: America's Romance with Wildlife on Film* (Cambridge, MA: Harvard Univ. Press, 1999).

6. See, for example, Roderick Nash, *Wilderness and the American Mind*, rev. ed. (New Haven: Yale Univ. Press, 1973); Peter J. Schmitt, *Back to Nature: The Arcadian Myth in Urban America* (New York: Oxford Univ. Press, 1969); Mark V.

Barrow, Jr., *A Passion for Birds: American Ornithology after Audubon* (Princeton, N.J.: Princeton Univ. Press, 1998).

7. Ralph H. Lutts, *The Nature Fakers: Wildlife, Science & Sentiment* (Golden, CO: Fulcrum, 1990).

8. The history of biology around the turn of the twentieth century is commonly told as the rise of laboratory research over outmoded natural history, and I have tried to resist this view. Lynn Nyhart challenges it using examples from Germany. See Lynn K. Nyhart, "Natural History and the 'New' Biology," in *Cultures of Natural History*, N. Jardine, J.A. Secord, and E.C. Spary (New York: Cambridge Univ. Press, 1996), pp. 426–443. On the history of zoological illustration during the transition from field to laboratory study, see Ann Shelby Blum, *Picturing Nature: American Nineteenth-Century Zoological Illustration* (Princeton, NJ: Princeton Univ. Press, 1993).

9. William T. Hornaday, "The Right Way to Teach Zoology," *The Outlook*, 1910, 95(June 4):256–263.

10. See, for example, John Kasson, *Amusing the Million: Coney Island at the Turn of the Century* (New York: Hill & Wang, 1978); Roy Rosenzweig, *Eight Hours for What We Will: Workers and Leisure in an Industrial City* (New York: Cambridge Univ. Press, 1983); Roy Rosenzweig and Elizabeth Blackmar, *The Park and the People: A History of Central Park* (Ithaca, NY: Cornell Univ. Press, 1992). On exhibits of people, see Robert W. Rydell, *All the World's a Fair* (Chicago: Univ. Chicago Press, 1984); Curtis Hinsley, "The World as Marketplace: Commodification of the Exotic at the World's Columbian Exposition, Chicago, 1893," in *Exhibiting Cultures: The Poetics and Politics of Museum Display*, ed. Ivan Karp and Steven D. Lavine (Washington, DC: Smithsonian Institution, 1991), pp. 344–365; Robert Bogdan, *Freak Show: Presenting Human Oddities for Amusement and Profit* (Chicago: Univ. Chicago Press, 1988).

11. "Spectacle" quote is from Michael Brambell, "The Evolution of the Modern Zoo," *International Zoo News*, 1993, 40(7):27; "zoology" quote is from William M. Mann, "A Visit to European Zoos," in *Explorations and Field Work of the Smithsonian in 1929* (Washington, DC: U.S. Government Printing Office, 1930), p. 62. Historians of science have begun to examine popular scientific practice, particularly popular natural history. See, for example, Elizabeth B. Keeney, *The Botanizers: Amateur Scientists in Nineteenth-Century America* (Chapel Hill, NC: Univ. North Carolina Press, 1992). Works that discuss the making of popular scientific texts include Dorothy Nelkin, *Selling Science: How the Press Covers Science and Technology* (New York: W.H. Freeman, 1987); Catherine A. Lutz and Jane L. Collins, *Reading National Geographic* (Chicago: Univ. Chicago Press, 1993). See also Rosenzweig and Blackmar, *The Park and the People*. For discussions of the problematic meaning of wilderness, middle-class meanings of nature, the interpretation of human-made landscapes as natural, and nature at

theme parks, see, for example, William Cronon, ed. *Uncommon Ground: Toward Reinventing Nature* (New York: W.W. Norton, 1995).

12. On zoos as emblems of imperialism, see Harriet Ritvo, "Exotic Captives," in *The Animal Estate: The English and Other Creatures in the Victorian Age* (Cambridge: Harvard Univ. Press, 1987); Keith Thomas, *Man and the Natural World: A History of the Modern Sensibility* (New York: Pantheon, 1983); Kay Anderson, "Culture and Nature at the Adelaide Zoo: New Frontiers in 'Human' Geography," *Transactions, Institute of British Geographers, NS*, 1995, 20:275–294; Helen L. Horowitz, "Animal and Man in the New York Zoological Park," *New York History*, 1975, 56: 426–455. See also John Berger, "Why Look at Animals?" in *About Looking* (New York: Pantheon, 1980); Bob Mullan and Garry Marvin, *Zoo Culture* (London: George Weidenfeld & Nicholson, 1987); Randy Malamud, *Reading Zoos: Representations of Animals and Captivity* (New York: New York Univ. Press, 1998).

13. This approach stems from the work of Michel Foucault. See, for example, Michel Foucault, *Power/Knowledge: Selected Interviews and Writings, 1972–1977* (New York: Pantheon, 1981) and *The Order of Things: An Archeology of the Human Sciences* (New York: Vintage, 1994). Tony Bennett follows Foucault in his efforts to "unravel the relations between knowledge and power effected by the technologies of vision embodied in the architectural forms of the exhibitionary complex." Tony Bennett, *The Birth of the Museum: History, Theory, Politics* (New York: Routledge, 1995), p. 63.

14. Carla Yanni argues a similar point in her insightful book on the ways in which the question "what is nature?" was answered through the architecture of Victorian natural history museums. She writes, "Some museums might present a single master narrative, but I argue that this is rare, and even if such a master narrative exists in one moment, it changes over time. There are usually several co-existing theories, rather than one master narrative, and the displays and architecture (if studied in precise historical detail) turn out to be surprisingly resistant to Foucauldian analysis. . . . Historians have put a lot of pressure on the meaning of 'orderliness' as a sign of social control, but do we as historians honestly think that curators ought to have arranged their collections in a disorderly fashion? Or, given that there are different kinds of order, a fashion which they believed to be disorderly?" Carla Yanni, *Nature's Museums: Victorian Science & the Architecture of Display* (Baltimore: Johns Hopkins Univ. Press, 1999), pp. 8–9.

15. Filson Young, "On Going to the Zoo," *Living Age*, 1912, 274(August 31):569.

16. The boundaries between environmental history, the history of natural history, and American culture have emerged as a rich area for historical inquiry. See, for example, Mitman, *Reel Nature*; Thomas R. Dunlap, *Saving America's Wildlife: Ecology and the American Mind, 1850–1990* (Princeton: Princeton

Univ. Press, 1988); Andrew C. Isenberg, *The Destruction of the Bison: An Environ-mental History, 1750–1920* (New York: Cambridge Univ. Press, 2000); Barrow, *A Passion for Birds*; Jennifer Price, *Flight Maps: Adventures with Nature in Modern America* (New York: Basic Books, 1999).

17. This book takes its source material mainly from municipal zoos that were founded around the turn of the twentieth century. But zoos, of course, are not the only places to see wild animals on display and they never have been. A central problem for the first American zoos was how to distinguish themselves from other animal shows, and zoos continue to struggle with this issue today. In addition to city zoos, there are dozens of privately run parks, ranging from roadside attractions to the Walt Disney Animal Kingdom near Orlando, Florida. Animal collections are displayed in theme parks as well, and drive-through safari parks offer a different way to see wildlife. About 160 of these organizations are accredited by the American Zoo and Aquarium Associ-ation, which sets standards for animal care and management and for profes-sional ethics, coordinates captive breeding programs, and promotes conserva-tion education. In addition, there are upwards of six hundred institutions licensed by the U.S. Department of Agriculture to exhibit wild animals in this country that might also be considered zoos. See Anthony D. Marshall, *Zoo: Profiles of 102 Zoos, Aquariums, and Wildlife Parks in the United States* (New York: Random House, 1994).

Chapter One
Animals in the Landscape

1. For a description of the Philadelphia Zoological Park in the late 1870s, see M. Howland, "The Philadelphia Zoo," *Harper's Magazine*, 1878, 58:699–712.

2. On the history of the Philadelphia Zoo, see The Zoological Society of Phil-adelphia, *An Animal Garden in Fairmount Park* (Philadelphia: The Zoological Society of Philadelphia, 1988); quotation is from Howland, "Philadelphia Zoo," p. 707.

3. See, for example, Robert and Gale McClung, "Early New Yorkers Liked Animal Shows," *Animal Kingdom*, 1957, 60 (3):88–93; quotation is from How-land, "Philadelphia Zoo," p. 709. Education and entertainment were the twin goals, similarly in tension, at public natural history museums and in national parks, and in wildlife films, in the same period. On films, see Gregg Mitman, *Reel Nature: America's Romance With Wildlife on Film* (Cambridge, MA: Harvard Univ. Press, 1999).

4. After P. T. Barnum's American Museum burned in 1865, Edwin Lawrence Godkin published an essay calling for a "real museum," modeled on the Brit-ish Museum, to replace it; Edwin Lawrence Godkin, "A Word About Muse-ums," *Nation*, 1865, 1(July 27):113–114. On museums, see Roy Rosenzweig and Elizabeth Blackmar, *The Park and the People: A History of Central Park* (Ithaca,

NY: Cornell Univ. Press, 1992), pp. 340–372; quotation is from Howland, "Philadelphia Zoo," p. 709; on middle-class landscapes, see Thomas J. Schlereth, "Chautauqua: A Middle Landscape of the Middle Class," *The Old Northwest*, 1986, 12:265–278.

5. See, for example, Gustave Loisel, *Histoire des Menageries de l'Antiquite a Nos Jours*, vols. 1–3 (Paris: Octave Doin et Fils, 1912); Lord Zuckerman, *Great Zoos of the World: Their Origins and Significance* (Boulder, CO: Westview Press, 1980); James Fisher, *Zoos of the World: The Story of Animals in Captivity* (Garden City: Natural History Press, 1967); R.J. Hoage and William A. Deiss, eds., *New Worlds, New Animals: From Menagerie to Zoological Park in the Nineteenth Century* (Baltimore: Johns Hopkins Univ. Press, 1996); Vicki Croke, *The Modern Ark: The Story of Zoos Past, Present and Future* (New York: Scribner, 1997); Kisling, ed., *Zoo and Aquarium History* (Boca Raton, FL: CRC Press, 2000); David Hancocks, *A Different Nature: The Paradoxical World of Zoos and Their Uncertain Future* (Berkeley and Los Angeles: Univ. California Press, 2001).

6. See Zuckerman, *Great Zoos*; the guidebook is C.V.A. Peel, *The Zoological Gardens of Europe: Their History and Chief Features* (London: F.E. Robinson, 1903); German zoos are also discussed in Lynn K. Nyhart, University of Wisconsin–Madison, "For Heimat and Empire: German Zoos, 1860–1880," unpublished manuscript.

7. The Amsterdam Zoo, *Artis*, was also an expression of nationalist sentiment, putting collections of animals and ethnographic objects from Dutch colonies on display. On Amsterdam, see Donna Christine Mehos, "Science Displayed: Nation and Nature at the Amsterdam Zoo *Artis*," Ph.D. Diss., University of Pennsylvania, 1997; quotations are from William T. Hornaday, "Report upon a Tour of Inspection of the Zoological Gardens of Europe," in New York Zoological Society, *First Annual Report*, 1897:35–36.

8. Quotations are from Zoological Society of Philadelphia, *Animal Garden*; see also Zoological Society of Philadelphia, *Second Annual Report*, 1873:27; final quotation is from Howland, "Philadelphia Zoo," p. 703.

9. On the Cincinnati Zoo, see David Ehrlinger, *The Cincinnati Zoo and Botanical Garden from Past to Present* (Cincinnati: Cincinnati Zoo and Botanical Garden, 1993), p. 5. Interestingly, before municipalities created public parks, Andrew Jackson Downing advocated the formation of joint stock companies to finance public gardens. See David Schuyler, *The New Urban Landscape: The Redefinition of City Form in Nineteenth-Century America* (Baltimore: Johns Hopkins Univ. Press, 1986), p. 55.

10. For examples of reports by Americans on European zoos, see New York Zoological Society, *First Annual Report*, 1897:35–43; Chicago Zoological Society, *Year Book*, 1927:51; on the history of pleasure grounds parks, see Galen Cranz, *The Politics of Park Design: A History of Urban Parks in America* (Cambridge, MA: MIT Press, 1982), pp. 3–59; on the perceived social threats of urbanization in general, see, for example, Paul Boyer, *Urban Masses and Moral Order in America*,

1820–1920 (Cambridge, MA: Harvard Univ. Press), 1978, and Robert Wiebe, *The Search for Order, 1877–1920* (New York: Hill and Wang, 1967); on the history of national parks, see, for example, Alfred Runte, *National Parks: The American Experience* (Lincoln: Univ. Nebraska Press, 1987); Richard West Sellars, *Preserving Nature in the National Parks: A History* (New Haven, CT: Yale Univ. Press, 1997); James A. Pritchard, *Preserving Yellowstone's Natural Conditions* (Lincoln: Univ. Nebraska Press, 1999).

11. The classic work on the idea of the middle landscape in American culture is Leo Marx, *The Machine in the Garden: Technology and the Pastoral Ideal in America* (New York: Oxford Univ. Press, 1964); the quotation is from Schlereth, "Chautauqua"; see also Donald Worster, *Nature's Economy: A History of Ecological Ideas*, 2nd ed. (New York: Cambridge Univ. Press, 1994).

12. On the relationship between urban Americans and nature, see, for example, Peter J. Schmitt, *Back to Nature: The Arcadian Myth in Urban America* (New York: Oxford Univ. Press, 1969); on the history of nature as a tourist attraction, see John F. Sears, *Sacred Places: American Tourist Attractions in the Nineteenth Century* (New York: Oxford Univ. Press, 1989) and Dona Brown, *Inventing New England: Regional Tourism in the Nineteenth Century* (Washington, DC: Smithsonian Institution, 1995); on the history of nineteenth-century American suburbs, see John Stilgoe, *Borderland: Origins of the American Suburb, 1820–1939* (New Haven, CT: Yale Univ. Press, 1988); quotations are from Sylvester Baxter, "The Trolley in Rural Parts," *Harpers New Monthly Magazine*, 1898, 97(June):61; see also Schlereth, "Chautauqua."

13. On Olmsted's parks as instruments of social control, see, for example, Geoffrey Blodgett, "Frederick Law Olmsted: Landscape Architecture as Conservative Reform," *Journal of American History*, 1976, 62:869–889.

14. See Cranz, *Politics of Park Design*, p. 24; quotations about Lincoln Park are from *Souvenir of Lincoln Park: An Illustrated and Descriptive Guide* (Chicago: Illinois Engraving Co., 1896), pp. 7,44; quotation about the Bronx Zoo is from William T. Hornaday, *Popular Official Guide to the New York Zoological Park as Far as Completed* (New York: New York Zoological Society, 1899), p. 28; more recently, landscaped parks, Olmsted's in particular, have often been mistaken for indigenous, untouched landscapes; see, for example, Susanna S. Zetzel, "The Garden in the Machine: The Construction of Nature in Olmsted's Central Park," in *Prospects: An Annual of American Cultural Studies*, Vol. 14, ed. Jack Salzman (New York: Cambridge Univ. Press, 1989), pp. 291–339, and Anne Whiston Spirn, "Constructing Nature: The Legacy of Frederick Law Olmsted," in *Uncommon Ground*, ed. William Cronon (New York: W.W. Norton, 1995), pp. 91–113.

15. On "enjoyment" see Cranz, *Politics of Park Design*, p. 14.

16. On privately run parks in general, see Rosenzweig and Blackmar, *The Park and the People*, pp. 104–111; for details on Augusta, see Colonel F.B. Dyer to Frank Baker, 6 Dec. 1901, Records of the National Zoological Park, RU 74,

box 100, folder 2, Smithsonian Institution Archives (SIA); on Silver Lake, see W.R. Lodge to Frank Baker, 20 Dec. 1901, RU 74, box 100, folder 2, SIA; on Boston, see Adams D. Claflin to Frank Baker, 27 Nov. 1901, RU 74, box 100, folder 3, SIA; on Longfellow Gardens, see R.F. Jones, *The Story of Longfellow Gardens* (Minneapolis: R.F. Jones, 1913); the quotation is from Baxter, "The Trolley in Rural Parts," p. 61.

17. On Downing's views of park use, see David Schuyler, *Apostle of Taste: Andrew Jackson Downing, 1815–1852* (Baltimore: Johns Hopkins Univ. Press, 1986), pp. 202–203. The desires of park planners often clashed with those of park users. On disputes over park activities and use see Cranz, *Politics of Park Design*, pp. 3–59; Rosenzweig and Blackmar, *The Park and the People*; Roy Rosenzweig, *Eight Hours for What We Will: Workers and Leisure in an Industrial City* (New York: Cambridge Univ. Press, 1992).

18. Frederick Law Olmsted, Sr., *Forty Years of Landscape Architecture: Central Park*, ed. Frederick Law Olmsted, Jr., and Theodora Kimball (reprint, Cambridge, MA: MIT Press, 1973), p. 514.

19. The quotations are from Boston Parks Department Annual Reports, 1901–1910, and Olmsted Brothers "1910 Report to the Boston Board of Park Commissioners," both quoted in Cynthia Zaitzevsky and Molly Gerard, "Franklin Park Draft Historic Landscape Report," June 10, 1987.

20. On circus menageries, see George L. Chindahl, *A History of the Circus in America* (Caldwell, ID: Caxton Printers, 1959); on nineteenth-century dime museums, see David Nasaw, *Going Out: The Rise and Fall of Public Amusements* (New York: Basic Books, 1993), pp. 10–18; the menagerie advertisements are quoted in Joanne Carol Joys, *The Wild Animal Trainer in America* (Boulder, CO: Pruett, 1983), p. 3; on Barnum's museum as a scientific institution, see John Rickards Betts, "P.T. Barnum and the Popularization of Natural History," *Journal of the History of Ideas*, 1959, 20:353–368.

21. On Amsterdam, see Gustave Loisel, "The Zoological Gardens and Establishments of Great Britain, Belgium, and the Netherlands," in Smithsonian Institution, *Annual Report*, 1907:442; on London, see New York Zoological Society, *Second Annual Report*, 1898:63; the "bewildering" quote is from William T. Hornaday, "The New York Zoological Park," *News Bulletin of the Zoological Society*, 1897, no. 1:3.

22. "Palace" quotation is from Hornaday, "The New York Zoological Park"; "pretentious" quotation is from "Notes by Mr. Frederick Law Olmsted, Jr.," July 31, 1902, RU 74, box 15, folder 11, SIA; "repair" quotation is from "Foreign Zoological Garden Notes," *News Bulletin of the New York Zoological Society*, No. 3, December 1898, 6; "fantastic" quotation is from F.W. True to Charles D. Walcott, Secretary of the Smithsonian Institution, 14 Nov. 1912, Records of the Office of the Secretary (Charles D. Walcott), 1903–1924, RU 45, box 87, SIA.

23. On Como Park in St. Paul, see *Como Zoo Guide* (St. Paul, MN: Como Zoo Volunteer Committee, 1960); on St. Louis, see Caroline Loughlin and Catherine

Anderson, *Forest Park* (St. Louis, MO: Junior League of St. Louis, 1986); on Swope Park, see Board of Park Commissioners, *Souvenir: The Park and Boulevard System of Kansas City* (Kansas City, MO: Board of Park Commissioners, 1914); on "Solitude," see *Art and Architecture at the Philadelphia Zoo* (Philadelphia: Zoological Society of Philadelphia, 1981, rev. 1988); the heights surrounding Cincinnati, in particular the suburb of Clifton, and their depiction in popular books, are discussed in Stilgoe, *Borderland*, pp. 56–64.

24. New York Zoological Society, *First Annual Report*, 1897: 14,15.

25. For a biography of Hornaday, see James Andrew Dolph, "Bringing Wildlife to Millions: William T. Hornaday, the Early Years, 1854–1896," Ph.D. Diss., University of Massachusetts, 1975; for a history of the relationships among the bison, Native American and Euroamerican hunters, and the ecology of the plains, see Andrew C. Isenberg, *The Destruction of the Bison* (New York: Cambridge Univ. Press, 2000).

26. On the history of the National Zoo, see Sybil E. Hamlet, "The National Zoological Park from Its Beginnings to 1973," typescript, Records of the National Zoological Park, Office of Public Affairs, box 37, folders 1–8, SIA; Helen L. Horowitz, "The National Zoological Park: 'City of Refuge' or Zoo?" *Records of the Columbia Historical Society of Washington, DC*, 1973–74, 49:405–429.

27. Quotations are from "The New York Zoological Society—Its Plans and Purposes," New York Zoological Society, *First Annual Report*, 1897:15; on the history of the Bronx Zoo, see also William Bridges, *Gathering of Animals: An Unconventional History of the New York Zoological Society* (New York: Harper & Row, 1974).

28. On the Bronx, see William T. Hornaday, "Report on the Character and Availability of South Bronx Park," in New York Zoological Society, *First Annual Report*, 1897:28,30; on Columbus, see Tod Raper, "How Columbus Built a Zoo," *Parks & Recreation*, 1940, 23(9):423–426; on Chicago, see *The Brookfield Zoo: 1934–1954*, (Brookfield, IL: Chicago Zoological Society, 1954); Hornaday is quoted in "Chicago's New Zoo: Dr. William T. Hornaday Visits City and Praises Site," *Parks & Recreation*, 1922, 5(6):580–582.

29. Hornaday was the author of *Free Rum on the Congo* (Chicago: Women's Temperance Publication Associations, 1887), protesting the traffic in rum to African natives. The quote about Michael Flynn is in Leonidas Hubbard, Jr., "What a Big Zoo Means to the People," *Outing*, 1904, 44:678; Hornaday's rubbish war is described in William T. Hornaday, "The Rubbish War," *Zoological Society Bulletin*, 1908, no. 31:456; arrests are described in New York Zoological Society, *Tenth Annual Report*, 1905; 44–45; "saloons" quote is from Madison Grant, "History of the Zoological Society," *Zoological Society Bulletin*, 1910, no. 37:597.

Reform was only one part of the cultural meaning of zoos, and I am not arguing that zoo audiences passively accepted zoos' reform rhetoric. Recent scholarship on the history of recreation and the history of public parks challenges a social control model; see, for example, Rosenzweig, *Eight Hours for What We Will*; Rosenzweig and Blackmar, *The Park and the People*.

30. The London Zoo song was written by Hugh Willoughby Sweny. See Lynne Iadarola, "Zoos," in *Encyclopedia of Architecture: Design, Engineering, and Construction*, Vol. 5 (New York: John Wiley, 1989), p. 424. On Hornaday's campaign, see *Zoological Society Bulletin*, 1910, no. 39:663; quotations are from William T. Hornaday, "Odious Nicknames," *Zoological Society Bulletin*, 1901, no. 5:24.

31. Quoted in William Bridges, *Gathering of Animals*, p. 116.

32. On Como Park in St. Paul, see *Como Zoo Guide*, quotations are from Olmsted, *Forty Years of Landscape Architecture* pp. 512, 509; on the Central Park menagerie, see Rosenzweig and Blackmar, *The Park and the People*, pp. 340–349.

33. On the Denver Zoo, see Carolyn Etter and Don Etter, *The Denver Zoo: A Centennial History* (Denver: Denver Zoological Foundation, 1995); on Baltimore, see Raymond Thompson, "Zoo, Second Oldest in America, Observing 75th Anniversary," Arthur Watson Library, Baltimore Zoo; on St. Louis, see Loughlin and Anderson, *Forest Park*; on Buffalo, see Roeder J. Kinkel, "New Zoological Garden Plans for Buffalo, N.Y.," *Parks and Recreation*, 1935, 19(4):131–134; on Atlanta, see Richard J. Reynolds III "History of the Atlanta Zoo," in *Atlanta's Zoo* (Atlanta: Eason Publications, 1969) and Francis Desiderio, "Raising the Bars: The Transformation of Atlanta's Zoo, 1889–2000," *Atlanta History*, 2000, 43(4):7–51; on Detroit, see William A. Austin, *The First Fifty Years: An Informal History of the Detroit Zoological Park* (Detroit: Detroit Zoological Society, 1974); on the AAZPA, see "Zoo Men Organize," *Parks and Recreation*, 1924, 8(2):121–122. In the 1990s the organization changed its name to the American Zoo and Aquarium Association.

34. See Thomas Schlereth, "Chautauqua," for a similar argument; on college campuses as a peculiarly American genre of landscape design, see Paul Venable Turner, *Campus: An American Planning Tradition* (Cambridge, MA: MIT Press, 1984).

35. Stilgoe, *Borderland*, pp. 52–55; Kenneth T. Jackson, *Crabgrass Frontier: The Suburbanization of the United States* (New York: Oxford Univ. Press, 1985), pp. 76–79; Witold Rybczynski, "How to Build a Suburb," *Wilson Quarterly*, 1995, 19(3):118–119.

36. On the view from the heights, see Stilgoe, *Borderland*, pp. 56–64.

37. On the Brookfield Zoo's architecture, see, for example, Chicago Zoological Society, *Year Book*, 1927; on Cincinnati, see Ehrlinger, *The Cincinnati Zoo*, p. 3; on the National Zoo, see Heather Ewing, "Architecture of the National Zoological Park," in *New Worlds, New Animals*; on the Franklin Park Zoo, see Zaitzevsky and Gerard, "Franklin Park Draft Historic Landscape Report"; on the Olmsteds and the Denver Zoo, see Etter and Etter, *The Denver Zoo*; on the

Olmsteds and Audubon Park, see L. Ronald Forman and Joseph Logsdon, *Audubon Park: An Urban Eden* (New Orleans: Friends of the Zoo, 1985).

38. On Chicago, see Chicago Zoological Society, *Year Book*, 1927; on Boston, see Arthur A. Shurtleff, "The Franklin Park 'Zoo,' Boston," *Architectural Review*, 1912 (March):29–31.

39. Mary N. Woods, "Thomas Jefferson and the University of Virginia: Planning the Academic Village," *Journal of the Society of Architectural Historians*, 1985, 44:266–283.

40. For a biography of Blackburne, see "Blackburne," typescript, Papers of William M. Mann and Lucile Quarry Mann, RU 7293, box 8, Smithsonian Institution Archives; on Robinson and Stephan, see Ehrlinger, *The Cincinnati Zoo*, pp. 19,23.

41. On the National Zoo and circus animals, see Sybil E. Hamlet, "The National Zoological Park from Its Beginnings to 1973"; on Clyde Beatty and the Detroit Zoo, see Austin, *The First Fifty Years*, p. 48; on Velox at the Denver Zoo, see Etter and Etter, *The Denver Zoo*, p. 5.

42. On Jo Mendi see Austin, *The First Fifty Years*, pp. 88–89; Detroit Zoological Park Commission, *Detroit Zoo-Life* (Detroit: Detroit Zoological Park Commission, 1932), p. 8.

43. On elephant rivalry, see Richard W. Flint, "American Showman and European Dealers," in *New Worlds, New Animals*, p. 103; on Coney Island attractions, see John F. Kasson, *Amusing the Million: Coney Island at the Turn of the Century* (New York: Hill & Wang, 1978), pp. 70–71.

44. On Buffalo, see Austin M. Fox, *Symbol and Show: The Pan-American Exposition of 1901* (Buffalo, NY: Meyer Enterprises, 1987); on St. Louis, see Murat Halstead, *Pictorial History of the Louisiana Purchase and World's Fair at St. Louis* (Philadelphia: Frank S. Brant, 1904); on Frank Buck, see *Official Guide Book, New York World's Fair, 1939* (New York: Exposition Publications, 1939), pp. 33, 41.

45. *A Century of Progress, Inc., Official Book of the Fair* (Chicago: Cuneo Press, 1932), p. 38.

46. On Ota Benga, see Samuel P. Verner, "The Story of Ota Benga, The Pygmy," *Zoological Society Bulletin*, 1916, 19(4):1377–1379; William T. Hornaday, "An African Pygmy," *Zoological Society Bulletin*, 1906, no. 23:301–302; Phillips Verner Bradford, *Ota: The Pygmy in the Zoo* (New York: St. Martin's, 1992); for references to Native Americans see Ehrlinger, *The Cincinnati Zoo*, p. 32 and Etter and Etter, *The Denver Zoo*, p. 28.

47. On amusement parks and fair midways, see Kasson, *Amusing the Million*; David Nasaw, *Going Out*.

48. Quotations are from F.W. True to Charles D. Walcott, Secretary of the Smithsonian Institution, 14 Nov. 1912, Records of the Office of the Secretary (Charles D. Walcott), 1903–1924, RU 45, box 87, SIA; the park manual referred

to is L.H. Weir, *Parks: A Manual of Municipal and County Parks*, 2 vols. (New York: A.S. Barnes and Company, 1928), p. 915.

49. The Bronx Zoo's Baird Court is discussed as a replica of the White City in Helen L. Horowitz, "Animal and Man in the New York Zoological Park," *New York History*, 1975, 56:426–455 and "Seeing Ourselves through the Bars: A Historical Tour of American Zoos," *Landscape*, 1981, 25(2):12–19.

50. The first quotation is from William T. Hornaday, "Report upon A Tour of Inspection of the Zoological Gardens of Europe," in New York Zoological Society, *First Annual Report*, 1897:36; the "civic spirit" quote is from "The New York Zoological Society—Its Plans and Purposes," New York Zoological Society, *First Annual Report*, 1897:13; "progress" quote is from New York Zoological Society, *Second Annual Report*, 1898:43; the Wichita Falls quote is from Jonnie R. Morgan to William M. Mann, 28 Nov. 1926, RU 74, box 97, folder 13, SIA.

51. "Highest power" quote is from Alpheus Hyatt, "The Next Stage in the Development of Public Parks," *Atlantic Monthly*, 1891, 67(400):215–224; on St. Louis, see "Public Fund Is Advocated for an Extension of Zoo," *St. Louis Post-Dispatch*, 28 Aug. 1910; on Evansville, see Gilmore M. Haynie, "Value and Problems of Zoological Parks in Smaller Cities," *Parks & Recreation*, 1930, 14(3):135–136; on Duluth, see "Zoo Reviews: The Zoological Gardens at Duluth, Minn." *Parks & Recreation*, 1933, 17(1):32; Memphis quote is in J.J. Williams to Frank Baker, 4 Nov. 1901, Records of the National Zoological Park, RU 74, box 100, folder 1, SIA. Frank Baker, who was Superintendent of the National Zoo, conducted a survey in 1901 to find out how many zoos existed in the United States, what animals they held, the number and size of their buildings, and the source of funds for their maintenance. He mailed his survey questions to the mayor of every city with a population of more than twenty-five thousand. The letter from Memphis is one of the responses he received; this correspondence is held in the SIA, RU 74, box 100, folders 1–3.

52. Sybil E. Hamlet, "The National Zoological Park from Its Beginnings to 1973," p. 10; for a comparison of New York zoo and museum endowment and attendance figures, see New York Zoological Society, *Fourteenth Annual Report*, 1909:108; on Philadelphia, see Zoological Society of Philadelphia, *Animal Garden*, pp. 14–15; on Cincinnati, see Ehrlinger, *The Cincinnati Zoo*, p. 33; on New York, see "Public School Visitors," *Zoological Society Bulletin* 1910, no. 41:696–698; on nature study and school trips at San Diego, see "Schoolwork in the Zoo," *San Diego Zoonooz*, 1931, 6(2):3–5; on San Diego's zoo science course, see H.C. Steinmetz, "Zoo Science," *Parks & Recreation*, 1933, 16(5):244–245.

53. For example, C.A. Failles, "The Value of a Zoo in a City Park," *Parks & Recreation*, 1918, 1(3):3–6; Howland, "Philadelphia Zoo." The quote referring to school children is from C. Emerson Brown, "A New Zoological Park for Vancouver," *Parks & Recreation*, 1932, 15(7):429.

Chapter Two
Who Belongs in the Zoo?

1. Susan Stewart's definition of a representative collection expresses well the intentions of zoo directors. She writes, "to have a representative collection is to have both the minimum and the complete number of elements necessary for an autonomous world—a world which is both full and singular, which has banished repetition and achieved authority." Susan Stewart, *On Longing: Narratives of the Miniature, the Gigantic, the Souvenir, the Collection* (Durham, NC: Duke Univ. Press, 1993), p. 152; the Hornaday quotation is from William T. Hornaday, *Popular Official Guide to the New York Zoological Park*, 1st ed. (New York: New York Zoological Society, 1899), p. 84; Belle J. Benchley, "We Give You Nature," *Parks & Recreation*, 1942, 26(2):87–88.

2. For example, Records of the National Zoological Park, RU 74, box 34, Smithsonian Institution Archives (SIA).

3. Typescript, "Views of Mr. Brown rel. NZP," 18 June 1890, RU 74, box 51, folder 10, SIA.

4. This correspondence is held in RU 74, boxes 34, 91, 92, and 97, SIA. It includes, in most cases, the zoo director's response. In addition, a single donation often generated four or more letters, and one donation could include more than one animal. So that the conclusions of this chapter may be taken to apply to American zoos in general, I have excluded letters about donations that were peculiar to the National Zoo because of its status as a government institution—gifts to the White House that were passed on to the zoo, for example. Recent studies of human-animal relationships provide guidance for thinking about how zoo audiences viewed wildlife, although these studies are generally not historical. On relationships between humans and animals that are pets, see James Serpell, *In the Company of Animals: A Study of Human-Animal Relationships* (New York: Basil Blackwell, 1986). Stephen Kellert breaks down late-twentieth-century human attitudes toward animals into ten categories. See, for example, Stephen R. Kellert, "The Biological Basis for Human Values of Nature," in *The Biophilia Hypothesis*, ed. Stephen R. Kellert and Edward O. Wilson (Washington, DC: Island Press, 1993), pp. 42–69.

5. A large and growing literature stresses the role of audiences not as passive consumers of entertainment, texts, or images but as active participants in shaping the meaning of mass-produced culture. The attempt here is to analyze the activities and interests of people who used the zoo. In contrast to the arguments made by historians of parks and other popular entertainments, zoos do not emerge as places for working-class rebellion against genteel standards of taste or behavior. See, for example, Roy Rosenzweig and Elizabeth Blackmar, *The Park and the People: A History of Central Park* (Ithaca, NY: Cornell Univ. Press, 1992); John F. Kasson, *Amusing the Million: Coney Island at the Turn of the Century* (New York: Hill & Wang, 1978); Kathy Peiss, *Cheap Amusements: Work-*

ing Women and Leisure in Turn-of-the-Century New York (Philadelphia: Temple Univ. Press, 1986); "AHR Forum: The Folklore of Industrial Society: Popular Culture and Its Audiences," *American Historical Review*, 1992, 97:1369–1430; Janice Radway, *Reading the Romance: Women, Patriarchy, and Popular Literature* (Chapel Hill: Univ. North Carolina Press, 1984).

6. "History of the Washington Park Zoo," *The Evening Wisconsin*, reprinted in *Souvenir of The Elephant Show*, February 12–13, 1906, box 26, Collection no. M89–188, Archives of the Zoological Society of Milwaukee County, State Historical Society of Wisconsin; William M. Mann, *Wild Animals In and Out of the Zoo*, Smithsonian Series, vol. 6 (1930; reprint New York: The Series Publishers, 1949), p. 4.

7. "A Boston 'Zoo'," *Science: A Weekly Newspaper of All the Arts and Sciences*, 1891, 17(417):64.

8. "A Boston 'Zoo'," *Science: A Weekly Newspaper of All the Acts and Sciences*, 1891, 17(417):63.

9. "A Boston 'Zoo'," *Science: A Weekly Newspaper of All the Acts and Sciences*, 1891, 17(417):63.

10. See, for example, Helen L. Horowitz, "The National Zoological Park: 'City of Refuge' or Zoo?" *Records of the Columbia Historical Society of Washington, D.C.*, 1973–74, 49:405–29; Horowitz, "Animal and Man in the New York Zoological Park," *New York History*, 1975, 56:426–55.

11. New York Zoological Society, *First Annual Report*, 1897:16.

12. Prairie dogs were equally as symbolic of the American west as bison, although they carried different connotations. See Susan Jones, "Becoming a Pest: Prairie Dog Ecology and the Human Economy in the Euroamerican West," *Environmental History*, 1999, 4(4):531–552.

13. William T. Hornaday, *Popular Official Guide to the New York Zoological Park*, 4th ed., 1901, p. 83.

14. Ned Hollister to Joseph T. Bethel, October 10, 1923, Records of the National Zoological Park, RU 74, box 101, folder 4, SIA.

15. Carolyn Etter and Don Etter, *The Denver Zoo: A Centennial History* (Denver: Denver Zoological Foundation, 1995), p. 31; Sybil E. Hamlet, "The National Zoological Park from Its Beginnings to 1973," Records of the National Zoological Park, Office of Public Affairs, RU 365, box 37, folder 1, p. 50.

16. The figure from the *Naturalists' Directory* is cited in Daniel Goldstein, "'Yours for Science': The Smithsonian Institution's Correspondents and the Shape of Scientific Community in Nineteenth Century America," *Isis*, 1994, 85:573–599; on amateur naturalists' motivations and practices, see also Elizabeth Barnaby Keeney, *The Botanizers: Amateur Scientists in Nineteenth-Century America* (Chapel Hill, NC: Univ. North Carolina Press, 1992); D.C. Beard, *The American Boys Handy Book* (1882; reprint, Lincoln, MA: David R. Godine, 1983).

17. On relationships between professional and amateur ornithologists, and the professionalization of ornithology, see Mark V. Barrow, Jr., *A Passion for*

Birds: American Ornithology after Audubon (Princeton, NJ: Princeton Univ. Press, 1998); for a prosopography of the Smithsonian's correspondents, and a revisionist profile of the American natural science community in the nineteenth century, see Daniel Goldstein, " 'Yours for Science' "; on Baird's network of collectors, see also E.F. Rivinus and E.M. Youssef, *Spencer Baird of the Smithsonian* (Washington, DC: Smithsonian Institution Press, 1990) and William Deiss, "Spencer F. Baird and His Collectors," *Journal of the Society for the Bibliography of Natural History*, 1980, 9:635–645.

18. Quotations are from the following sources: boy collectors, William H. Babcock to Frank Baker, 7 Sept. 1893, RU 74, box 34, folder 5, SIA; F.H. McHaffie to National Zoological Gardens, 27 Sept. 1923, RU 74, box 91, folder 8, SIA; bats, Mrs. J.H. Cummings to William M. Mann, 30 June 1927, RU 74, box 91, folder 11, SIA; owls, unsigned letter, received 20 July 1907, RU 74, box 91, folder 6, SIA.

19. Frank Baker, "The National Zoological Park," in *The Smithsonian Institution 1846–1896, The History of Its First Half Century*, ed. George Brown Goode (Washington, DC: Smithsonian Institution, 1897), p. 454.

20. On relationships between museums and collectors, and collecting guides, see Barrow, *A Passion for Birds*, pp. 20–145.

21. See *Animals Desired for the National Zoological Park at Washington D.C.* (Washington, DC: Government Printing Office, 1899); curassow, Charles H. Allen to Superintendent of the Zoo, 29 July 1901, RU 74, box 91, folder 4, SIA; quote from *Animals Desired*, p. 4; on the Venezuelan consul, see E.H. Plumacher to Hon. David J. Hill, 2 Oct. 1899, RU 74, box 34, folder 1, SIA.

22. Thomas Barbour's gift of two spotted turtles was acknowledged in Smithsonian Institution, *Annual Report*, 1927:94; on his donation of a Central American tapir to the Philadelphia Zoo, see Zoological Society of Philadelphia, *56th Annual Report*, 1928:33; on Edmund Heller, see Roger Conant, "Edmund Heller Passes Away," *Parks & Recreation*, 1939, 22(12):638–639; on J. Alden Loring, see Etter and Etter, *The Denver Zoo*, pp. 47–50.

23. St. Louis newspaper clipping, St. Louis Zoological Park Records, 1910–1941, SL 194, unidentified clipping in scrapbook, vol. 1, Western Historical Manuscripts Collection, University of Missouri–St. Louis; donations are listed in, for example, Secretary's Annual Report (typescript), 7 April 1919, SL 194, folder 2.

24. Quotation is from G.W. Armistead to Smithsonian Museum, 15 Feb. 1897, RU 74, box 34, folder 4, SIA; on turn-of-the-century dime museums, see David Nasaw, *Going Out: The Rise and Fall of Public Amusements* (New York: Basic Books, 1993).

25. On Barnum and others, see John Rickards Betts, "PT Barnum and the Popularization of Natural History," *Journal of the History of Ideas*, 1959, 20:353–368; chicken, H.C. Baird to Smithsonian Institute, 30 Nov. 1895, RU 74, box 34,

folder 5, SIA; rooster, M.D. Oakley to Superintendent, National Zoological Park, 15 May 1901, RU 74, box 97, folder 7, SIA.

26. Quotations are from the following sources: hairless mare, A. Frank Moore to Manager Zoo Park, 25 Sept. 1900, RU 74, box 97, folder 6, SIA; goat, Mrs. Nicholas Fox to Manager, 31 April 1901, RU 74, box 97, folder 7, SIA; horned sheep, W.R.C. Johnstone to Smithsonian Institution, 10 Aug. 1901, RU 74, box 97, folder 7, SIA; six-legged dog, H.H. Hern to Smithsonian Institute, 27 May 1907, RU 74, box 97, folder 10, SIA; earless hog, R.T. Swann to Smithsonian, 17 Oct. 1901, RU 74, box 97, folder 7, SIA; greatest wonder, W.H. Davis to Zoological Garden, 10 April 1905, RU 74, box 97, folder 10, SIA.

27. Frank Baker to A.H. Childs, 11 July 1905, RU 75, box 97, folder 10, SIA; see also Frank Baker to J.W. Prude, 13 Sept. 1900, RU 74, box 97, folder 6, SIA.

28. Parrot, Elizabeth A. Arnold to Frank Baker, 4 Aug. 1899, RU 74, box 34, folder 4, SIA; Coast Guard bear, James Philip Schneider to William M. Mann, 17 May 1928,RU 74, box 91, folder 3, SIA; monkey, Mary B.Y. Wynkoop to William M. Mann, 3 Sept. 1928, and WMM to MBYW, 5 Sept. 1928, RU 74, box 91, folder 3, SIA.

29. Pigeons, Frank Baker to Clinton P. Townsend, 26 June 1903, RU 74, box 91, folder 2, SIA; parrot, E.N. Johnston to Frank Baker, 13 Dec. 1915 and William Blackburne memo, RU 74, box 91, folder 4, SIA.

30. On the St. Louis camel, see newspaper clipping, St. Louis Zoological Park Records, 1910–1941, SL 194, unidentified clipping in scrapbook, vol. 1, Western Historical Manuscripts Collection, University of Missouri–St. Louis; donations are listed in, for example, Secretary's Annual Report (typescript), 7 April 1919, SL 194, folder 2; Shriners, unidentified newspaper article, SL 194, vol. 1; Philadelphia, Zoological Society of Philadelphia, *57th Annual Report*, 1929:34; military mascot, for example, typed note, RU 74, box 91, folder 6, SIA.

31. Eagles, H.J. Fallon to Ned Hollister, 11 Jan. 1923, RU 74, box 91, folder 8, SIA; parrot, Mrs. C.M. Buck to C.W. Walcott, 7 Sept. 1923, RU 74, box 91, folder 8, SIA.

32. Parrot, Mrs. C.M. Buck to C.W. Walcott, 7 Sept. 1923, RU 74, box 91, folder 8, SIA; coyote, Clyde Howard to National Zoological Park, 20 March 1925, RU 74, box 91, folder 9, SIA; fox, F.G. Shaible to Ernest Seton-Thompson, 17 May 1900, RU 74, box 91, folder 4, SIA; wild cat, J.A. August to Frank Baker, 23 Dec. 1897 and 2 Jan. 1898, RU 74, box 34, folder 4, SIA.

33. Peary's sled dogs, see Zoological Society of Philadelphia, "An Animal Garden in Fairmount Park," 1988, typescript, p. 15; Denver wolf, Frances Melrose, "Black Jack, the Wolf," *Rocky Mountain News*, April 18, 1982, p. 4; burro, The Tolman Laundry to Frank Baker, 5 Aug. 1904, RU 74, box 91, folder 6, SIA; goat, C.A. Beard to William Blackburne, 3 Dec. 1895, RU 74, box 34, folder 11, SIA; Chinese bird, Genevieve B. Wimsatt to William M. Mann, 18 Nov. 1925 and undated, RU 74, box 91, folder 9, SIA.

34. Monkey, Mrs. Robert B. Stiles to Manager of Zoo, 5 April 1924, RU 74, box 91, folder 8 and 18 May 1925, RU 74, box 91, folder 9, SIA; opossum, Fannie M. Hawkins to Superintendent, 4 Feb. 1923, RU 74, box 91, folder 8, SIA; baboon, H.N. Slater to Ned Hollister, 16 April 1921, RU 74, box 91, folder 6, SIA.

35. See John E. Lodge, "How Jungle Beasts Live Near Our Big Cities," *Popular Science Monthly*, 1932, 121(October):26–27,104; Wilson Chamberlain, "Backstage at the Zoo," *Scientific American*, 1937, 156(May):290–291; Freeman M. Shelly, "Cage Service," *Parks & Recreation*, 1939, 23(2):68–74. Gorilla diet and Waldorf-Astoria quotes in "How Jungle Beasts Live," pp. 27,104; strangest hotel quote in "Cage Service," p. 68. Some people expected zoo-animal celebrities to make public appearances outside the zoo. Edward S. Schmid, for example, wrote to the National Zoo to ask whether a camel (and keeper) could be borrowed to march in a parade for a Masonic temple; the zoo director refused. See Edward S. Schmid to Frank Baker, 3 May 1916, RU 74, box 91, folder 3, SIA.

36. Japanese robin, G.R. Brigham to Mr. Hollister, 5 Jan. 1921, RU 74, box 91, folder 7, SIA; Belle Benchley quoted in Belle J. Benchley, "Why a Zoo?" *Parks & Recreation*, 1932, 15(5):285. Belle Benchley was the first female director of an American zoo. She had joined the San Diego zoo staff as secretary to the director in 1926. Her popular books include Belle Benchley, *My Life in a Man-Made Jungle* (Boston: Little, Brown, 1940), and *My Friends the Apes* (Boston: Little, Brown, 1942).

37. White cat, Miss E. Fish to Frank Baker, Oct. 1906, RU 74, box 91, folder 6, SIA.

38. Campaign results, Boston *Post*, 7 June 1914, p. 8; on the history of pageants and their cultural function, see David Glassberg, *American Historical Pageantry: The Uses of Tradition in the Early Twentieth Century* (Chapel Hill: Univ. North Carolina Press, 1990); the "play together" motto was coined by the St. Louis Pageant Drama Association in 1914; see Glassberg, *American Historical Pageantry*, p. 177.

39. On Atlanta see Francis Desiderio, "Raising the Bars: The Transformation of Atlanta's Zoo, 1889–2000," *Atlanta History*, 2000, 43(4):7–51 and Richard J. Reynolds III, "History of the Atlanta Zoo," in *Atlanta's Zoo*, ed. Deborah Eason and Elton Eason (Atlanta: Eason Publications, 1969), p. 15; on Baltimore, see Raymond Thompson, "Zoo, Second Oldest in America, Celebrating 75th Anniversary," undated manuscript in Baltimore Zoo library; on St. Louis, see St. Louis *Post-Dispatch*, undated, SL 194, vol. 1; on New Orleans, see Ronald Forman, *Audubon Park: An Urban Eden* (New Orleans: Friends of the Zoo, 1985), p. 128; on Evansville, IN, see Gilmore M. Haynie, "Value and Problems of Zoological Parks in Smaller Cities," *Parks & Recreation*, 1930, 14(3):136; for a discussion of a newspaper campaign to raise money for children bitten by rabid dogs, see Bert Hansen, "America's First Medical Breakthrough: How Popular

Excitement about a French Rabies Cure in 1885 Raised New Expectations for Medical Progress," *American Historical Review*, 1998, 103(2):373–418.

40. On the zoo opening, see *Souvenir Guide to Franklin Park Zoo and Botanical Gardens* (Boston: Franklin Park, 1951), p. 2; on the history of the zoo, see Richard Heath, "A History of the Franklin Park Zoo," *Jamaica Plain Citizen*, June 25–August 6, 1981; animals are listed in Boston *Post*, March 13, 1914; on buffalo, see *Post*, December 8, 1913.

41. Boston *Post*, 6 Mar. 1914; *Post*, 9 Mar. 1914.

42. Boston *Post*, 9 Mar. 1914; quote, *Post*, 11 Mar. 1914.

43. Testimonials, Boston *Post*, 9 Mar. 1914; SPCA, *Post*, 10 Mar. 1914.

44. Elephant editor, Boston *Post*, 9 March 1914; business contributors included the B.F. Keith Theater, where the elephants had been performing, movie theaters, and department stores. Barber shops and soda fountains set up collection boxes in their stores. See *Post*, 19 Mar. 1914; 5 Apr. 1914; 11 Mar. 1914.

45. "Sweet memories," Boston *Post*, 24 Mar. 1914; "love children," *Post*, 11 Mar. 1914.

46. Post office, Boston *Post*, 16 Mar. 1914; hockey team, *Post*, 18 Mar. 1914; riding school, *Post*, 22 Mar. 1914; jury, *Post*, 24 Mar. 1914; Ancients, *Post*, 27 Mar. 1914; assessors and penal department, *Post*, 28 Mar. 1914; Sharon train, *Post*, 7 Apr. 1914.

47. "Bulk" quote, Boston *Post*, 9 Mar. 1914; cartoon, *Post*, 31 Mar. 1914; "Visiting day," Boston *Post*, 23 Mar. 1914.

48. "No sum," *Boston Post*, 13 Mar. 1914.

49. Rat catcher, Boston *Post*, 16 Mar. 1914; "stars and hundreds," *Post*, 19 Mar. 1914; Sarah Wilson, *Post*, 25 Mar. 1914; Mary Flynn, *Post*, 16 Mar. 1914; J. Costello, *Post*, 2 Apr. 1914; A. Schwarzenberg, *Post*, 11 Apr. 1914; photos, *Post*, 28 Mar. 1914.

50. B.F. Keith's Boston theater was one of the lavish vaudeville "palaces" built in the 1890s to attract middle-class audiences. Top prices for seats in these theaters ran as high as $1.50, compared to twenty-five cents in smaller theaters. For a description of Keith's and an interpretation of vaudeville as respectable entertainment, see David Nasaw, "Something for Everybody," in *Going Out*, pp. 19–33.

51. "Burmese," Boston *Post*, 6 Mar. 1914; "ruling princes," *Post*, 9 Mar. 1914; "African," *Post*, 20 Mar. 1914; 200 years old, *Post*, 9 Mar. 1914. Confusion about the lifespan of elephants evidently was widespread. According to one zoo guidebook of the era, "The ordinary life of the elephant is supposed to be about a hundred years, although in special cases they undoubtedly live much longer." In Arthur Erwin Brown, *Guide to the Garden of the Zoological Society of Philadelphia*, 8th ed., 1900, pp. 42–43.

52. On Jumbo, see Harriet Ritvo, *The Animal Estate: The English and Other Creatures in the Victorian Age* (Cambridge, MA: Harvard Univ. Press, 1987), p.

232; "while the boys and girls," Boston *Post*, 3 Apr. 1914; "little English boy," *Post*, 16 Mar. 1914.

53. "Best example," Boston *Post*, 30 Mar. 1914; Tony finds dime, *Post*, 17 Mar. 1914; "a little slower," *Post*, 9 Mar. 1914.

54. "Three trick elephants," "News From Zoos," *Parks & Recreation*, 1939, 22(8):428; on Dunk and Gold Dust, see Sybil E. Hamlet, "The National Zoological Park From Its Beginnings to 1973."

55. "Clever Pet," for example, Boston *Post*, 26 Mar. 1914; collie, *Post*, 14 Mar. 1914; canary, *Post*, 16 Mar. 1914; Nemo and Dixie, *Post*, 22 Mar. 1914; Roxbury kitten, *Post*, 30 Mar. 1914.

56. Ceremony, Boston *Post*, 7 June 1914; "regular Bostonians," *Post*, 8 June 1914.

57. Quote is from "History of the Washington Park Zoo."

58. Harriet Ritvo writes that for visitors to feed the animals in the London Zoo "was an act which symbolized both proprietorship and domination," and that elephant and camel rides "encouraged zoo visitors to think of [animals] as temporary possessions or playthings." Ritvo, *The Animal Estate*, p. 220. In contrast, the sense of ownership communicated through the elephant campaign seems like a longer-term commitment, and a more complicated relationship, than Ritvo describes.

59. On the ways in which souvenirs mediate time and space, see Stewart, *On Longing*, pp. 132–151.

60. Another segment of the zoo-going public, for example, took pleasure from teasing and torturing the animals, and feeding them sharp objects or poisons. As the authority on zoo management Heine Hediger put it, "In all ages and places menageries and zoos have attracted not only men with a healthy interest but all kinds of abnormal ones upon whom the zoo seems to act like a magnet." He describes these offenders in Heine Hediger, *Wild Animals in Captivity* (New York: Dover, 1964), pp. 173–176.

Chapter Three
The Wild Animal Trade

1. On Ellis Joseph's platypus cage, see "A Great Animal Collector," *Parks & Recreation*, 1922, 6(2):147–149; quotation is from the label for the platypus display, a photograph of which is printed in William T. Hornaday, "Observations on Zoological Park Foundations," *Zoological Society Bulletin*, 1925, 28(1):16.

2. Natural history dealers, trading in taxidermied animals, bird eggs, minerals, shells, insects, fossils, and so on, also became established in the United States after the Civil War. See Mark V. Barrow, Jr., "The Specimen Dealer: Entrepreneurial Natural History in America's Gilded Age," *Journal of the History of Biology*, 2000, 33:493–534.

3. The terms "collector" and "dealer" distinguish mainly between people who went out into the field to collect and people who stayed in their offices selling animals collected by others. This boundary, however, was fluid; some collectors sold their animals to dealers as well as directly to zoos and circuses, and dealers sometimes went into the field. A few zoos did mount their own expeditions; the Bronx Zoo was among the first to do so. Expeditions to collect animals for the National Zoo are the subject of Chapter 4 of this book.

4. On the history of dealers in natural history specimens, see Mark V. Barrow, Jr., "The Specimen Dealer." The animals collected might be thought of as boundary objects—objects of common interest to zoos, circuses, collectors, and hunters, but for different reasons. Star and Griesemer use this concept to show how scientific work results from the cooperation of many different actors with varying viewpoints—"the functioning of mixed economies of information with different values and only partially overlapping coin." They use as their case study the collection of museum specimens, but the idea applies equally well to zoo collecting. See Susan Leigh Star and James R. Griesemer, "Institutional Ecology, 'Translations' and Boundary Objects: Amateurs and Professionals in Berkeley's Museum of Vertebrate Zoology, 1907–39," *Social Studies of Science*, 1989, 19:387–420. The rhinoceros delivered to the Bronx Zoo was immortalized in sculpture, which stands today outside the elephant house.

5. On the Dutch East India Company, and the early animal trade in general, see Richard W. Flint, "American Showmen and European Dealers: The Commerce in Wild Animals in 19th Century America," in *New Worlds, New Animals: From Menagerie to Zoological Park in the Nineteenth Century*, ed. R.J. Hoage and W.A. Deiss (Baltimore, MD: Johns Hopkins Univ. Press, 1996), pp. 97–108; see also George L. Chindahl, *A History of the Circus in America* (Caldwell, ID: Caxton, 1959), p. 2.

6. P. Chalmers Mitchell, "Introduction," in *Beasts and Men*, by Carl Hagenbeck, abridged translation by Huge S.R. Elliot and A.G. Thacker (London: Longmans, Green, 1909), p. vi.

7. This story is recounted in Hagenbeck's autobiography, Hagenbeck, *Beasts and Men*, pp. 1–7; a thorough and thoughtful analysis of the Hagenbeck company history is provided in Nigel Rothfels, *Savages and Beasts: The Birth of the Modern Zoo* (Baltimore, MD: Johns Hopkins Univ. Press, 2002).

8. Hagenbeck, *Beasts and Men*, p. 12.

9. The caravan evidently traveled by rail rather than using the newly opened Suez Canal. Hagenbeck, *Beasts and Men*, pp. 12–14; on the importance of this shipment to Hagenbeck's career, see Rothfels, *Savages and Beasts*.

10. The quotation about "every German consul" appears in Frank Buck with Ferrin Fraser, *All in a Lifetime* (New York: Robert M. McBride, 1941), p. 113; on the Prjwalsky's horses, which were collected at the request of the Duke of Bedford for his private park, see Hagenbeck, *Beasts and Men*, p. 74; Hagenbeck quote, *Beasts and Men*, p. 11; import figures, Herman Reichenbach, "Carl Ha-

genbeck's Tierpark and Modern Zoological Gardens," *Journal of the Society for the Bibliography of Natural History*, 1980, 9(4):574; on Hagenbeck in relation to other German businesses, see Rothfels, *Savages and Beasts*.

11. David Ehrlinger, *The Cincinnati Zoo and Botanical Garden from Past to Present* (Cincinnati: Cincinnati Zoo and Botanical Garden, 1993), p. 42.

12. On the London dealers, see Harriet Ritvo, *The Animal Estate: The English and Other Creatures in the Victorian Age* (Cambridge, MA: Harvard Univ. Press, 1987), pp. 244–245; on the Reiche brothers, see Flint, "American Showmen and European Dealers," pp. 101–105; on Van Amburgh, see Wilton Eckley, *The American Circus* (Boston: Twayne, 1984), p. 51; the Records of the National Zoological Park (RU 74) at the Smithsonian Institution Archives contain correspondence and price lists pertaining to Ruhe, Bartels, Trefflich, Mohr, and Foehl.

13. For a partial list of American circuses and menageries to 1956, see Chindahl, *A History of the Circus in America*, pp. 240–271; for a list of wild animal acts in America, see also Joanne Carol Joys, *The Wild Animal Trainer in America* (Boulder, CO: Pruett, 1983), pp. 296–297; on Barnum's visit to Hagenbeck and Hagenbeck's elephant exports, see Hagenbeck, *Beasts and Men*, pp. 11, 29; on elephant envy see Flint, "American Showmen and European Dealers," p. 106; see also Chindahl, *A History of the Circus in America*, p. 104.

14. Quotations are from an advertisement for Court's act in Eckley, *The American Circus*, pp. 58–59; on animal trainers, see Eckley, *The American Circus*, pp. 51–81. Clyde Beatty was also known for his books, including Clyde Beatty, *The Big Cage* (New York: The Century Co., 1933), and *Jungle Performers* (New York: R.M. McBride, 1941), and his movies *The Big Cage* (1933) and *The Lost Jungle* (1934), both produced by Universal Studios.

15. On the preference for wild-caught animals, see Eckley, *The American Circus*, p. 73; the Hagenbeck and I.S. Horne price lists are located in the Records of the National Zoological Park, RU 74, box 114, folder 10, Smithsonian Institution Archives (SIA).

16. For animal prices, see, for example, price lists in RU 74, boxes 157–165, SIA; on the Sparks Circus caravan, see Eckley, *The American Circus*, p. 40.

17. On Jumbo, see, for example, Ritvo, *The Animal Estate*, p. 232; on Cincinnati, see Ehrlinger, *The Cincinnati Zoo*, p. 14; on the National Zoo, see Sybil E. Hamlet, "The National Zoological Park from Its Beginnings to 1973," typescript, Records of the National Zoological Park, Office of Public Affairs, RU 365, box 37, folders 1–8, p. 34, SIA.

18. William A. Conklin, *Report of the Director of the Central Park Menagerie* (New York: William C. Bryant, 1873), p. 6.

19. The correspondence quoted is William Blackburne to Frank Baker, 19 Dec. 1890, RU 74, box 34, SIA.

20. On Colonel E. Daniel Boone, see correspondence, Boone to Frank Baker, RU 74, box 143, folder 4, SIA; on Stephan, see, for example Ehrlinger, *The Cincinnati Zoo*, p. 14.

21. On Bostock as competition for the Baltimore Zoo, see Raymond Thompson, "Zoo, Second Oldest in America, Observing 75th Anniversary," undated typescript, located in the Baltimore Zoo library. Bostock was British, and a descendant of George Wombwell, famous as the first English menagerie owner; see Joys, *The Wild Animal Trainer in America*, p. 25. On Bostock at the Pan-American Exposition, see correspondence, Frank C. Bostock to Frank Baker, RU 74, box 63, folders 3 and 4, SIA; on Conklin, see correspondence, William Conklin to Frank Baker, RU 74, box 35, folder 11, SIA; on Reynolds, see "Returns to Benson," *Parks & Recreation*, 1936, 19(5):170.

22. Pamphlet advertisement for Chas. Payne, RU 74, box 63, folder 1, SIA; between 1890 and 1940 the different directors of the National Zoo corresponded with hundreds of animal collectors and dealers.

23. William M. Mann, who became director of the National Zoo after working for years as a field entomologist, recalled in his autobiography that he read Beard's book as a boy. See William M. Mann, *Ant Hill Odyssey* (Boston: Little, Brown, 1948) p. 8. Daniel C. Beard was an artist and popular writer of nature stories and boy books before he became the founder of the American Boy Scouts. See Daniel C. Beard, *The American Boys Handy Book: What to Do and How to Do It* (New York: Scribners, 1882,1890; reprint, Boston: Godine, 1983). A companion book for girls, written by Daniel Beard's sisters, makes no mention of collecting birds or taxidermy, but rather teaches skills appropriate for young female naturalists: how to preserve wild flowers, for example. See Lina Beard and Adelia B. Beard, *The American Girls Handy Book: How to Amuse Yourself and Others* (New York: Scribners, 1887,1898; reprint, Boston: Godine, 1987).

Some collectors' wives assisted their husbands in taking care of animals, but few were collectors or dealers themselves. One exception was a "Miss O'Duffy" who had an animal-dealing business in New York City around 1890; after she married the director of the Central Park Menagerie in 1890 or 1891, the couple managed the business together. See correspondence in RU 74, box 35, folder 11, SIA. Another female animal dealer was Genevieve Cuprys, also known as "Jungle Jenny," who was the foster daughter of the dealer Arthur Foehl, and made trips to India, Singapore, and Siam in the late 1940s. See "J is for Jungle Jenny," in Henry Trefflich, as told to Baynard Kendrick, *They Never Talk Back* (New York: Appleton-Century-Crofts, 1954), pp. 113–136.

24. Biographies of naturalists who grew up in the late nineteenth century typically include a passage in which the aspiring zoologist's parents urge him to set his birds and snakes free, pay more attention to school, and pursue a career in, say, medicine or the military. See, for example, Robert Henry Welker, *Natural Man: The Life of William Beebe* (Bloomington, IN: Indiana Univ. Press, 1975), and L.N. Wood, *Raymond L. Ditmars: His Exciting Career with Reptiles, Animals and Insects* (New York: Julian Messner, 1944).

25. "Odd Ways of Making a Living," *The Museum* (Albion, NY), 1896, 2(8):42.

26. On Ward, his business, and its role as a scientific supply house, see Sally Gregory Kohlstedt, "Henry A. Ward: The Merchant Naturalist and American Museum Development," *Journal of the Society for the Bibliography of Natural History*, 1980, 9:647–661; Ward and other natural history entrepreneurs are discussed in Barrow, "The Specimen Dealer."

27. See correspondence between Caraway and Frank Baker, RU 74, box 63, folder 1, SIA.

28. The "set determination" quotation is from Frank B. Armstrong, "Notes From Interior of Mexico," *The Museum*, 1894, 1(1):22–24, on p. 23. The second quote is from Armstrong's stationery; see Armstrong to Frank Baker, 1902, RU 74, box 94, folder 7, SIA; quote from zoo director is from Frank Baker to James H. Morgan, 27 May 1905, RU 74, box 100, folder 7, SIA; for the Snake King reference, see, for example, unpaginated advertisement, *Parks & Recreation*, 1927, 10(6).

29. On Johnson, see price list in RU 74, box 114, folder 10, SIA; on Odell, see correspondence between W. Odell and Frank Baker, 1902, RU 74, box 94, folder 7, SIA.

30. Biographical information on Mayer is from Charles Mayer, *Trapping Wild Animals in Malay Jungles* (New York: Duffield, 1922); see also Charles Mayer, *Jungle Beasts I Have Captured* (Garden City, NY: Doubleday, 1924).

31. Quote is from Lamb's stationery, Chester A. Lamb to Edw. Schmid, 26 Apr. 1902, RU 74, box 63, folder 6, SIA; zoo director quote is in Frank Baker to James H. Morgan, RU 74, box 100, folder 7, SIA; on Ansel Robison see for example, price lists in RU 74, box 94, folder 10 and RU 74, box 115, folder 1, SIA.

32. On restrictions on animal imports and on the import of African animals in particular, see Edward H. McKinley, *The Lure of Africa: American Interests in Tropical Africa, 1919–1939* (New York: Bobbs-Merrill, 1974), pp. 117–147; on gifts of animals to zoos, see Chapter 2 of this book.

33. On San Simeon see William W. Murray to National Zoological Park, 29 Oct. 1940, RU 74, box 186, folder 4, SIA; on Percy Godwin, see Harry M. Wegeforth, *It Began with a Roar: The Beginnings of the World-Famous San Diego Zoo* (1953; reprint, San Diego: Zoological Society of San Diego, 1990), p. 107; for biographical information on Victor Evans, who frequently donated animals to the National Zoo, see William M. Mann, "Death of Victor Justice Evans," *Parks & Recreation*, 1931, 14(9):519–520; on Beebe's pheasant expedition see William Beebe, *A Monograph on the Pheasants*, 4 vols. (London: Witherby, 1918–1922); on Thorne, see William T. Hornaday to Joel W. Thorne, 9 May 1924, letterpress volume 51 of Hornaday's outgoing correspondence, letter no. 684, New York Zoological Park, Wildlife Conservation Society Archives. Hornaday suggested that Thorne have the dealer Louis Ruhe sell his zebras on commission. On Hagenbeck's rhesus monkey sale, see Rothfels, *Savages and Beasts*. The rhesus trade grew rapidly. In each of 1936, 1937, and 1938, twelve thousand

rhesus monkeys were imported into the United States. In 1937, the British colonial government in India placed restrictions on rhesus exports because of humane society complaints regarding the conditions under which the monkeys were transported. See Clarence Ray Carpenter, "Rhesus Monkeys (*Macaca mulatta*) for American Laboratories," *Science*, 1940, 92:284–286.

34. On Bascom, see correspondence, 1938, RU 74, box 164, folder 10, SIA; on Bird Wonderland, see the price list in RU 74, box 115, folder 1, SIA; on Ross Allen see correspondence, 1937–1956, and "Men Who Make Florida," unidentified newspaper clipping, RU 74, box 157, folder 5, SIA.

35. For a summary of Benson's career, see Fletcher A. Reynolds, "Biographical Sketch of John T. Benson," *Parks & Recreation*, 1948, 31(11):655–666.

36. On I.S. Horne, see correspondence between Horne and the National Zoo director, RU 74, box 18, folder 4; RU 74, box 114, folder 10; RU 74, box 159, folder 4; SIA. On Bartels see correspondence and price lists, RU 74, box 115, folder 1; RU 74, box 157, folder 10, SIA. On Trefflich, see Henry Trefflich, *They Never Talk Back*, and correspondence and price lists, RU 74, box 164, folder 12; RU 74, box 186, folder 4, SIA. On Meems Brothers and Ward, see correspondence and price lists, RU 74, box 161, folder 6; and RU 74, box 186, folder 2, SIA.

37. On Joseph turning away biographers, see "Ellis Joseph Dies," *Parks & Recreation*, 1938, 22(2):72–73; on J.L. Buck, see J.L. Buck, "The Chimpanzee Shaken out of His Nest," *Asia*, 1927, 27:308–313,326,328; the books about Sasha Siemel are Julian Duguid, *Green Hell* (New York: Appleton-Century, 1931) and *Tiger Man* (New York: Appleton-Century, 1932). Siemel later published his memoirs, Sasha Siemel, *Tigrero* (New York: Prentice-Hall, 1953).

38. On Andrews, Akeley, and other naturalist-explorers for the American Museum of Natural History, see, for example, Douglas J. Preston, *Dinosaurs in the Attic: An Excursion into the American Museum of Natural History* (New York: St. Martin's, 1986); on the Johnsons, see Gregg Mitman, *Reel Nature: America's Romance With Wildlife on Film* (Cambridge, MA: Harvard Univ. Press, 1999), pp. 26–35, and Pascal James Imperato and Eleanor M. Imperato, *They Married Adventure: The Wandering Lives of Martin and Osa Johnson* (New Brunswick, NJ: Rutgers Univ. Press, 1992).

39. Quotation and biographical information are from Buck's autobiography, Buck with Fraser, *All in a Lifetime*; quote is on pp. 86–87.

40. Buck with Fraser, *All in a Lifetime*, pp. 114–115,254–256,274–275; Buck's other books are Frank Buck with Edward Anthony, *Bring 'Em Back Alive* (New York: Simon and Schuster, 1930); Frank Buck with Edward Anthony, *Wild Cargo* (New York: Simon and Schuster, 1932; reprint New York: Lancer, 1966); Frank Buck, *Capturing Wild Elephants!* (Chicago: Merrill, 1934); Frank Buck with Ferrin Fraser, *Fang and Claw* (New York: Simon and Schuster, 1935); Frank Buck with Ferrin Fraser, *Tim Thompson in the Jungle* (New York: Appleton-

Century, 1935); Frank Buck with Ferrin Fraser, *On Jungle Trails* (Yonkers, NY: Stokes World Book Co., 1936); Frank Buck with Ferrin Fraser, *Jungle Animals* (New York: Random House, 1945). Films that Buck produced and starred in include *Bring 'Em Back Alive* (1932); *Wild Cargo* (1934); *Fang and Claw* (1935); *Jungle Menace* (1937); and *Jacare—Killer of the Amazon* (1942). On Frank Buck in St. Paul, see *Como Zoo Guide* (St. Paul: Como Zoo Volunteer Committee, 1960), p. 1; on Buck in Baltimore, see Arthur R. Watson, "A Reptile House for Baltimore," *Parks & Recreation*, 1948, 31(10):597–598.

41. Buck with Anthony, *Wild Cargo*, p. 201.

42. For a gendered analysis of the collection of dead specimens, see Donna Haraway, "Teddy Bear Patriarchy: Taxidermy in the Garden of Eden, New York City, 1908–36," in *Primate Visions: Gender, Race, and Nature in the World of Modern Science* (New York: Routledge, 1989), pp. 26–58. On the sportsman's code, see, for example, John F. Reiger, *American Sportsmen and the Origins of Conservation* (New York: Winchester Press, 1975), pp. 25–49; the rules of "manly sport with the rifle" are also clearly articulated in the constitution of the elite hunting club, the Boone and Crockett Club. See "Constitution," in *Hunting and Conservation: The Book of the Boone and Crockett Club*, ed. George Bird Grinnell and Charles Sheldon (New Haven, CT: Yale Univ. Press, 1925), pp. 535–536.

43. Buck with Anthony, *Wild Cargo*, p. 157.

44. Quotation is from Buck with Anthony, *Wild Cargo*, p. 98.

45. Nigel Rothfels has analyzed the popular books of Carl Hagenbeck's animal catchers. See Rothfels, *Savages and Beasts*; Buck with Anthony, *Bring 'Em Back Alive*, pp. 48–60.

46. E. L. Doctorow, *World's Fair* (New York: Random House, 1985), pp. 262–263. Frank Buck's stories continue to have an audience in the twenty-first century, in a new edition published in 2000; *Frank Buck, Bring 'Em Back Alive: The Best of Frank Buck*, ed. Steven Lehrer (Lubbock: Texas Tech Univ. Press, 2000).

47. On zoo directors' opinions of Frank Buck, see Bernard Livingston, *Zoo Animals, People, Places* (New York: Arbor House, 1974), pp. 174–176.

48. See correspondence between Ennio Arrigutti and William M. Mann, 1939–1941, RU 74, box 157, folder 8, SIA.

49. On Hagenbeck's caravan, see Rothfels, "Bring 'Em Back Alive," p. 96; on Frank Buck's pheasants, see Buck with Fraser, *All in a Lifetime*, pp. 108–112; on the rhinos, see Buck with Anthony, *Bring 'Em Back Alive*, pp. 48–90.

50. The first quotation from Mitchell is in Hagenbeck, *Beasts and Men*, p. vi; the second quote is from *Beasts and Men*, p. v.; on Hagenbeck's experiments in acclimatization, see *Beasts and Men*, pp. 202–219; Hagenbeck quote is from *Beasts and Men*, p. 202.

51. Frank Baker to Arthur E. Brown, 30 Oct. 1908, RU 74, box 18, folder 4, SIA; an early book on keeping wild animals was R.B. Sanyal, *Handbook of the*

Management of Animals in Captivity (1892), but it was not widely distributed; the first general guide to zoo management, both practical and philosophical, was Heini Hediger, *Wild Animals in Captivity*, 1st English ed., trans. Geoffrey Sircom (London: Butterworth, 1950); the first comprehensive book on keeping specific species, still widely used, was L. Crandall, *Management of Wild Mammals in Captivity* (Chicago: Univ. Chicago Press, 1964).

52. From an advertisement for the Cincinnati Zoo, in *History of the Cincinnati Fire Department* (Cincinnati, OH: Firemen's Protective Association, 1895).

Chapter Four
Zoo Expeditions

1. As director of the National Zoo, Mann also made collecting trips to Cuba and Central America (1930), British Guiana (1931), and the Argentine Republic (1939), but these trips were shorter, less expensive, and less elaborate than the expeditions described here. Other zoos had sent collecting parties after particular animal species, but none had mounted an expedition on the scale of the one Chrysler sponsored.

2. The quotation from Mann is in "Zoological Expedition," undated typescript, RU 7293, Papers of William M. Mann and Lucile Quarry Mann, box 6, Smithsonian Institution Archives (SIA).

3. On zoo tours as safaris, see, for example, "Spring Safari at New York," *Parks & Recreation*, 1939, 22(11):578–582; it could also be argued that zoos promoted a kind of vicarious imperialism.

4. See Philip Pauly, "The World and All That Is in It: The National Geographic Society, 1888–1918," *American Quarterly*, 1979, 31:517–532.

5. For insight into how local conditions, in particular tourism and colonial culture, shape expedition work, this chapter is indebted to Alex Soojung-Kim Pang, "The Social Event of the Season: Solar Eclipse Expeditions and Victorian Culture," *Isis*, 1993, 84:252–277; Pang's arguments are developed more fully in Alex Soojung-Kim Pang, *Empire and the Sun: Victorian Solar Eclipse Expeditions* (Stanford: Stanford Univ. Press, 2002).

Scholarship on the construction of popular science has focused on the editorial negotiations involved in producing popular scientific texts. See, for example, Catherine A. Lutz and Jane L. Collins, *Reading National Geographic* (Chicago: Univ. Chicago Press, 1993); Greg Myers, *Writing Biology: Texts in the Social Construction of Scientific Knowledge* (Madison: Univ. Wisconsin Press, 1990); Dorothy Nelkin, *Selling Science: How the Press Covers Science and Technology* (New York: W.H. Freeman, 1987).

This chapter follows Bruno Latour's broad definition of scientific practice to include field work as well as popularization among the activities important in making scientific knowledge. See Bruno Latour, *Science in Action: How to Follow*

Scientists and Engineers Through Society (Cambridge, MA: Harvard Univ. Press, 1987). Historians of science have begun to characterize the practice of collecting dead specimens for natural history museums. For example, see David E. Allen, *The Naturalist in Britain: A Social History* (London: Allen Lane, 1976; reprint, Princeton, NJ: Princeton Univ. Press, 1994); Susan L. Star, "Craft vs. Commodity, Mess vs. Transcendence: How the Right Tool Became the Wrong One in the Case of Taxidermy and Natural History," in *The Right Tools for the Job*, ed. Adele E. Clarke and Joan H. Fujimura (Princeton, NJ: Princeton Univ. Press, 1992); Anne L. Larsen, "Not Since Noah: The English Scientific Zoologists and the Craft of Collecting, 1800–1840," Ph.D. Diss., Princeton University, 1993. The practice of popular collecting is the subject of Anne Secord, "Science in the Pub: Artisan Botanists in Early Nineteenth-Century Lancashire," *History of Science*, 1994, 32:269–315.

6. Biographical information is drawn from press releases, correspondence, and obituaries found in Records of the National Zoological Park, Office of Public Affairs, RU 365, box 39, SIA; colorful reminiscences about Mann's early field work are found in William M. Mann, *Ant Hill Odyssey* (Boston: Little, Brown, 1948); "small in stature" quote is from Thomas E. Snyder, John E. Graf, and Marion R. Smith, "William M. Mann, 1886–1960," *Proceedings of the Entomological Society Washington*, 1963, 63(1):70; the popular book quoted is Gordon MacCreagh, *White Waters and Black* (New York: The Century Co., 1926), p. 75; the expedition was known formally as the Mulford Biological Exploration of the Amazon Basin, and was sponsored by the H.K. Mulford Company, with H.H. Rusby appointed expedition director.

7. On Mann's ant collection, see press release, 27 Oct. 1956, RU 365, box 39, SIA; on the zoo's collection when Mann took over, see Sybil E. Hamlet, "The National Zoological Park from Its Beginnings to 1973," typescript, RU 365, box 37, folder 1, p. 135, SIA; the "representative collection" quotation is from an undated typescript, "Zoological Expedition," RU 7293, box 6, SIA; Mann never elaborated on what he meant by a "representative collection," but Susan Stewart, in an essay on collections, offers an interesting definition and one that seems appropriate to zoos: "to have a representative collection is to have both the minimum and the complete number of elements necessary for an autonomous world—a world which is both full and singular, which has banished repetition and achieved authority." See Susan Stewart, *On Longing: Narratives of the Miniature, the Gigantic, the Souvenir, the Collection* (Baltimore: Johns Hopkins Univ. Press, 1984), p. 152.

8. On Frank Buck and the Dallas Zoo, see Frank Buck with Edward Anthony, *Bring 'Em Back Alive* (New York: Simon and Schuster, 1930), p. 51; as Mann recounted his hiring at the zoo, "Before the matter of entrusting me with the Zoo came up the Secretary [of the Smithsonian] asked me the personal question 'Are you ready to settle down?' I answered it with an honest 'No.'

See undated script of a radio broadcast, one of several on 'Experiences of Dr. Mann in Africa,' located in RU 7293, box 12, SIA.

9. Loisel's book remains the most comprehensive general history of zoos: Gustave Loisel, *Histoire des Menageries de l'Antiquite a Nos Jours*, vol. 3 (Paris: Octave Doin et Fils, 1912), p. 331. The quotation from Hornaday is in a letter: William Temple Hornaday to Mr. Joel W. Thorne, 9 May 1924, W.T. Hornaday outgoing correspondence, letterpress vol. 51, no. 684, New York Zoological Society (NYZS) The Wildlife Conservation Society Archives.

10. On commercial dealers, see Chapter 3 of this book.

11. The quotation from Mann on his desire to maintain wide contacts is from the undated script of a radio broadcast, one of several on "Experiences of Dr. Mann in Africa."

12. William M. Mann to Theodore MacManus, 7 Jan. 1926, RU 7293, box 6, SIA; on popular films about African wildlife in the first decades of the twentieth century, see Gregg Mitman, *Reel Nature: America's Romance with Wildlife on Film* (Cambridge, MA: Harvard Univ. Press, 1999).

13. William M. Mann to Theodore MacManus, 1 Mar. 1926, RU 7293, box 6, SIA.

14. The quotation is from George Eastman, *Chronicles of an African Trip* (Rochester, NY: John P. Smith, 1927), p. 15. Eastman described the passengers aboard the ship. On the convergence of the four expeditions, see also Edward H. McKinley, *The Lure of Africa* (New York: Bobbs-Merrill, 1974), p. 117.

15. The many ways in which African wildlife was symbolic of the continent to Americans are discussed in depth in McKinley, *The Lure of Africa*, pp. 117–147; Mary Jobe Akeley, the second wife of taxidermist Carl Akeley wrote that "The idea that Africa is the 'world's zoo' is indelibly impressed upon the general mind," in Mary Jobe Akeley, *Carl Akeley's Africa* (New York: Dodd, Mead, 1929), p. 106; Carl Akeley's African Hall at the American Museum of Natural History was conceived as a monument to Roosevelt. McKinley, in *The Lure of Africa*, also discusses the 1909 Roosevelt expedition; on this expedition, see also Smithsonian Institution, *Annual Report*, 1910:43,101; Roosevelt's books about the 1909 expedition are Theodore Roosevelt, *African Game Trails: An Account of the African Wanderings of an American Hunter-Naturalist* (New York: C. Scribner's Sons, 1910), and Theodore Roosevelt and Edmund Heller, *Life Histories of African Game Animals* (New York: C. Scribner's Sons, 1914). Mann's "collected over" quotation is from a typescript prepared for a series of radio broadcasts, "Experiences of Dr. Mann in Africa."

16. The quotation from the Secretary of the Smithsonian is in Charles D. Walcott to Walter Chrysler, 13 Feb. 1926, RU 46, Records of the Office of the Secretary, 1925–1949, box 147, SIA; the quotation from William Mann is from a typescript prepared for a series of radio broadcasts, "Experiences of Dr. Mann in Africa."

More than fifty letters survive from people eager to join the expedition; they are located in RU 74, box 194, SIA. Most of the writers described themselves as young and able bodied, and willing to accept the thrill of adventure as the only compensation for their work.

17. Instances in which the Field Museum, the Cleveland Museum of Natural History, and the Los Angeles Museum of Natural History gained specimens by lending their names to big-game hunters are discussed in McKinley, *The Lure of Africa*, pp. 131–132. Mann identifies Carnochan in William M. Mann to Rodgers D. Hamilton, 14 Oct. 1946, RU 7293, box 1, SIA.

18. On African game conservation, see John M. MacKenzie, *The Empire of Nature: Hunting, Conservation and British Imperialism* (New York: Manchester Univ. Press, 1988); hunting licenses in British-ruled Africa are discussed in McKinley, *The Lure of Africa*, pp. 123–126; on Loveridge, see Sybil E. Hamlet, "The National Zoological Park from Its Beginnings to 1973"; the quotation is from a draft of a letter from the Secretary of the Smithsonian to the Secretary of State, Feb. 1926, RU 7293, box 6, SIA.

19. The Marine Corps supplies are detailed in William M. Mann to Office of the Post Quartermaster, 8 Mar. 1926, RU 74, box 194, SIA.

20. William M. Mann to Mr. Getz and Mr. Clark, 12 Jul. 1926, RU 7293, box 12, SIA.

21. Mann's description of hiring Goss, and the quotation, are from a typescript prepared for a series of radio broadcasts, "Experiences of Dr. Mann in Africa."

22. Mann's account of corralling giraffe is in a typescript prepared for a series of radio broadcasts, "Experiences of Dr. Mann in Africa."

23. On animal numbers and giraffe names, see Sybil E. Hamlet, "The National Zoological Park from Its Beginnings to 1973"; on the Smithsonian endowment, see William M. Mann to Theodore MacManus, 7 Jan. 1926, RU 7293, box 6, SIA; on the Science Service, see the clipping of the Science Service advertisement "Seven Great Scientists" in RU 365, box 45, scrapbook 1924–1931, SIA.

24. On the new zoo buildings, see Sybil E. Hamlet, "The National Zoological Park from Its Beginnings to 1973"; the quotation is from William M. Mann to Alexander Wetmore, 16 May 1937, RU 46, box 144, SIA.

25. Lucile Q. Mann was a 1918 graduate of the University of Michigan, and had worked as assistant editor at the Bureau of Entomology of the U.S. Department of Agriculture from 1918 to 1922, and as an editor at *The Women's Home Companion* in New York City from 1922 to 1926. Although she gave up full-time employment when she married, Lucile Mann continued to write, basing many of her articles and books on her experiences as wife of the zoo director. She also published a book on home aquaria. She became the primary chronicler of her husband's animal collecting expeditions, and later was an editor in the National Zoological Park administrative offices, from 1951 until her retirement in 1967. Biographical information on Lucile Mann is drawn from the

guide to RU 7293, SIA. Lucile Mann's books are Lucile Q. Mann, *From Jungle to Zoo: Adventures of a Naturalist's Wife* (New York: Dodd, Mead & Co., 1934); *Friendly Animals, A Book of Unusual Pets* (New York: Leisure League of America, 1935); and *Tropical Fish* (New York: Leisure League of America, 1935; rev., ed., New York: Sentinel Books, 1943, 1947).

The Manns describe the American animals accompanying the expedition in William M. Mann and Lucile Q. Mann, "Around the World for Animals," *National Geographic*, 1938, 73(June):666.

26. Mann evidently had the opportunity to speak to Grosvenor informally. He submitted his expedition proposal, "Apropos of our conversation the other evening"; see William M. Mann to Gilbert Grosvenor, 19 Oct. 1934, folder 11–10015.363, Research Grants–Mann, National Geographic Society Records Library, Washington, DC. Mann may have met Grosvenor through an article he published in *National Geographic* magazine the same year; see William M. Mann, "Stalking Ants, Savage and Civilized," *National Geographic*, 1934, 66:171–192. Grosvenor is quoted in a letter supporting Mann's request for collecting permits in the East Indies: Gilbert Grosvenor to Jonkheer H.M. van Haersma de With, 19 Nov. 1936, RU 7293, box 5, SIA; a copy of the "Memorandum of Agreement" between the National Geographic Society and William M. Mann is located in RU 7293, box 5, SIA.

27. Transportation in the colony is described in the *1930 Handbook of the Netherlands East Indies* (Buitenzorg, Java: Division of Commerce of the Department of Agriculture, Industry and Commerce); on the history of the trade in natural objects, see, for example, Wilfred T. Neill, *Twentieth-Century Indonesia* (New York: Columbia University Press, 1973), pp. 82–83.

28. For a discussion of German botanists at Buitenzorg, see Eugene Cittadino, *Nature as the Laboratory: Darwinian Plant Ecology in the German Empire, 1880–1900* (New York: Cambridge Univ. Press, 1990); on Hornaday's collecting, see William T. Hornaday, *Two Years in the Jungle: The Experiences of Hunter and Naturalist in India, Ceylon, the Malay Peninsula and Borneo* (New York: Charles Scribner's Sons, 1886); the commercial animal trade is discussed in detail in Chapter 3 of this book.

29. Quotations are from William M. Mann to William M. Wheeler, 21 Feb. 1933, RU 7293, box 5, SIA; Lucile Mann describes socializing with the other expeditions in her diary; see Lucile Q. Mann diary, 1937, RU 7293, box 7, folder 1, SIA; on Coolidge's expedition, see, for example, Harold J. Coolidge, Jr., "Trailing the Gibbon to Learn about Man," *The New York Times Magazine*, August 1, 1937, pp. 6–7+.

30. For descriptions of nature reserves, see *1930 Handbook of the Netherlands East Indies*, p. 31; American museum collectors, organized by scientists at Harvard's Museum of Comparative Zoology, supported the movement among colonial governments to protect wildlife, and formed the American Committee for International Wildlife Protection in 1930. See John P. Droege. "Specimen

Collectors to Conservationists: Scientists and International Conservation Before the Second World War," presented at the History of Science Society Annual Meeting, October 13–16, 1994, New Orleans.

31. The animals Mann desired are listed in William M. Mann to Walter A. Foote, American consul in Batavia (undated copy), RU 7293, box 5, SIA; letters of introduction are in RU 7293, box 5, SIA.

32. Harry M. Wegeforth to William M. Mann, 26 Aug. 1936, RU 7293, box 3, SIA.

33. Lucile Q. Mann diary, 1937, RU 7293, box 7, folder 1, SIA.

34. William T. Hornaday to J.J. White, 6 May 1924, outgoing correspondence, Director's office, New York Zoological Park, letterpress vol. 51, no. 677, NYZS, The Wildlife Conservation Society Archives.

35. Lucile Q. Mann diary, 1937, RU 7293, box 7, folder 1, SIA.

36. William M. Mann to Thomas W. McKnew, 31 July 1937, folder 11–3.132, Research Grants–Mann, National Geographic Society Records Library.

37. William M. Mann to Alexander Wetmore, 23 Mar. 1937, RU 46, box 144, SIA.

38. Lucile Q. Mann diary, 1937, RU 7293, box 7, folder 1, SIA.

39. For discussions of culture brokers, see Leroy Vail, ed., *The Creation of Tribalism in Southern Africa* (Berkeley: University of California Press, 1989); and Lynette L. Schumaker, "The Lion in the Path: Fieldwork and Culture in the History of the Rhodes–Livingstone Institute, 1937–1964," Ph.D. Diss., University of Pennsylvania, 1994.

40. For an example of colonial caricatures of the Dyak, see Carl Bock, *The Head-Hunters of Borneo* (London: Sampson Low, Marston, Searle & Rivington, 1881; reprint, New York: Oxford Univ. Press, 1985). Biographical information about Gaddi is drawn from W.H. Shippen, Jr., "Borneo Guide, Here, Is Friend of King," *The Evening Star*, 1 Oct. 1937; RU 7225, Smithsonian Institution Manuscript Collections, Anthony J. Conway Collection, circa 1929–1970, box 1, folder 5, SIA; William M. Mann to Layang Gaddi (undated copy), RU 7293, box 5, SIA; and William M. Mann to Thomas W. McKnew, 14 Aug. 1937, folder 502–1.1213, Research Grants–Mann, National Geographic Society Records Library. On Y. Siah, see J. Holbrook Chapman to William M. Mann, 29 Mar. 1937, RU 7293, box 5, SIA. For a discussion of Wallace's relationship with Ali, see Jane Camerini, "Wallace in the Field," *Osiris*, 2nd ser., 1996, 11:44–65.

41. William M. Mann to Alexander Wetmore, 26 May 1937, RU 46, box 144, SIA.

42. Calvin Daly to William M. Mann, 18 Jan. 1939, RU 74, box 252, folder 10, SIA.

43. Few colonial zoos have been written about. One example is the zoo in Colombo, Ceylon, run by John Hagenbeck, half-brother of the German animal dealer Carl Hagenbeck. John Hagenbeck wrote several books about his experi-

ences. The history of the zoo in Calcutta is recounted in D.K. Mittra, "Ram Bramha Sanyal and the Establishment of the Calcutta Zoological Gardens," in *New Worlds, New Animals: From Menagerie to Zoological Park in the Nineteenth Century*, ed. R.J. Hoage and William A. Deiss (Baltimore: Johns Hopkins Univ. Press, 1996), pp. 86–96. The quotations are from Lucile Q. Mann diary, 1937, RU 7293, box 7, folder 1, SIA.

44. Lucile Q. Mann diary, 1937, RU 7293, box 7, folder 1, SIA. Commercial dealers often lost an even higher percentage of their animals during transport.

45. Charles Morrow Wilson, *Liberia* (New York: William Sloane Associates, 1947); Yekutiel Gershoni, *Black Colonialism: The Americo-Liberian Scramble for the Hinterland* (Boulder, CO: Westview Press, 1985).

46. See Gershoni, *Black Colonialism*; D. Elwood Dunn, *A History of the Episcopal Church in Liberia, 1821–1980* (Metuchen, NJ: Scarecrow Press, 1992); D. Elwood Dunn and Svend E. Holsoe, *Historical Dictionary of Liberia: African Historical Dictionaries, No. 38* (Metuchen, NJ: Scarecrow Press, 1985); Wilson, *Liberia*, pp. 133–134.

47. Harold J. Coolidge, Jr., who was just finishing his undergraduate degree, was a member of the expedition. The report of the Harvard expedition was published as *The African Republic of Liberia and the Belgian Congo, Based on the Observations Made and Material Collected during the Harvard African Expedition, 1926–27* (Cambridge, MA: Harvard Univ. Press, 1930); the animal collector Hagenbeck sent to Liberia recounted his adventures in Hans Schomburgk, "On the Trail of the Pygmy Hippo," *Zoological Society Bulletin*, 1912, 16:880–884.

48. These events are recounted in Sybil E. Hamlet, "The National Zoological Park from Its Beginnings to 1973."

49. It is not clear how the Liberian animals were displayed at the New York World's Fair. The Firestone exhibit was in the "Transportation" area. In one guide book, published before the Liberian animals arrived at the fair, the Firestone exhibit is described as including a "life-size reproduction of a thriving American farm" featuring domestic animals and "the adoption of pneumatic tires to every kind of wheeled farm implement." See the *Official Guide Book: New York World's Fair, 1939* (New York: Exposition Publications, 1939), p. 165. Regarding travelogues of Liberia, see Graham Greene, *Journey without Maps* (Garden City, NY: Doubleday, 1936); and Susan L. Blake, "Travel and Literature: The Liberian Narratives of Esther Warner and Graham Greene," *Research in African Literatures*, 1991, 22:191–203.

During 1980 and 1981, Lucile Mann narrated the 16 mm film for oral historian Pamela M. Henson. This film has been transferred to videotape. The audiotaped narration has been synchronized with the image, and is kept in RU 7293, boxes 26 and 27, SIA.

50. Lucile Q. Mann diary, 1940, RU 7293, box 7, folder 4, SIA.

51. Lucile Q. Mann diary, 1940, RU 7293, box 7, folder 4, SIA.

52. The quotation from William Mann is in William M. Mann to Alexander Wetmore, 8 Apr. 1940, RU 46, box 144, SIA; traps used in Liberia are discussed in Wilson, *Liberia*, pp. 51–58; Lucile Mann is quoted from Lucile Q. Mann diary, 1940, RU 7293, box 7, folder 4, SIA.

53. See Lucile Q. Mann diary, 1940, RU 7293, box 7, folder 4, SIA.

54. Lucile Mann recounts this adventure in Lucile Q. Mann diary, 1940, RU 7293, box 7, folder 4, SIA.

55. Lucile Q. Mann diary, 1940, RU 7293, box 7, folder 4, SIA.

56. Lucile Q. Mann diary, 1940, RU 7293, box 7, folder 4, SIA.

57. On the pygmy hippo, see Lucile Q. Mann diary, 1940, RU 7293, box 7, folder 4, SIA; on species named for Firestone, see William M. Mann to Harvey Firestone, Jr., 21 Jan. 1943, RU 74, box 194, folder 4, SIA.

58. Lucile Q. Mann diary, 1940, RU 7293, box 7, folder 4, SIA.

59. On cutting vegetables, see Lucile Q. Mann diary, 1940, RU 7293, box 7, folder 4, SIA; for the list of animals, see RU 74, box 194, folder 4, SIA.

60. Directed by John Ford, *Mogambo* also starred Grace Kelly and Ava Gardner.

61. See William M. Mann, "Smithsonian–Chrysler Expedition to Africa to Collect Living Animals, Field Season of 1926," in *Explorations and Field-Work of the Smithsonian Institution in 1926*, pp. 10–21; William M. Mann and Lucile Q. Mann, "Collecting Live Animals in Liberia," in *Explorations and Field-Work of the Smithsonian Institution in 1940*, pp. 13–20; for an example of a write-up on dead specimens, see Arthur Loveridge, "Report on the Smithsonian-Firestone Expedition's Collection of Reptiles and Amphibians from Liberia," *Proceedings of the United States National Museum*, 1941, 91:113–140.

62. Lucile Q. Mann diary, 1937, RU 7293, box 7, folder 1, SIA.

63. William M. Mann to Thomas W. McKnew, August 14, 1937, folder 502–1.1213, Research Grants–Mann, National Geographic Society Records Library.

64. Maynard Owen Williams to Thomas W. McKnew, August 8, 1937, folder 502–1.1213, Research Grants–Mann, National Geographic Society Records Library. *National Geographic* was not the only media outlet to portray Gaddi as a primitive head hunter. Lucile Mann noticed the irony in his situation; she recounted NBC's live, national broadcast from the ship's deck, when the expedition landed in Boston, in her diary. Reporters interviewed Gaddi: "[William Mann] introduced Gaddi, who said a few precise, well-chosen words in Dyak. The announcer in describing Gaddi, said he was . . . in his native Dyak costume (this consisted of a blue beret, an imitation mackinaw, a pair of dirty pants, a knife in his belt, and tennis shoes.)" See Lucile Q. Mann diary, 1937, RU 7293, box 7, folder 1, SIA.

65. Maynard Owen Williams to Thomas W. McKnew, July 8, 1937, folder 502–1.1213, Research Grants–Mann, National Geographic Society Records Library.

66. The newspaper headline is from the Akron *Beacon-Journal*, 3 Nov. 1940; Mann related the Roosevelt anecdote, for example, in William M. Mann to George Seybold, 6 Mar. 1941, RU 7293, box 3, SIA.

67. Smithsonian Institution press release, RU 46, box 144, SIA.

68. In *Writing Biology*, Greg Myers compares examples of popular and professional writing by the same scientist, and argues that popular science writing tells "narratives of nature" in which plants or animals are the subject, as opposed to the "narratives of science" told in professional science journals, in which scientific activities are the subject. Popular accounts of Mann's expeditions suggest that narratives of culture—encounters with exotic peoples—are important in popularizing field work. There is no professional, technical counterpart to Mann's popular writing, but it would be interesting to see whether other field workers focus on narratives of culture in their popular writing, and to compare popular and technical accounts of the same expedition's work. See Myers, "The Social Construction of Popular Science: The Narrative of Science and the Narrative of Nature," in *Writing Biology*, pp. 141–192.

69. See Pauly, "The World and All That Is in It."

Chapter Five
Natural Settings

1. The first and third quotations are from "A Prairie Scene in New York," *Zoological Society Bulletin*, 1911, no. 47:801; the second quote is from William T. Hornaday, "Preliminary Plan for the Prosecution of the Work of the Zoological Society," Correspondence, Director's Office, New York Zoological Park, box 2, folder 9, New York Zoological Society (NYZS), The Wildlife Conservation Society Archives.

2. Clyde E. Hill, "What Price—Zoo Success?" *Parks & Recreation*, 1928, 11(5):387.

3. On the history and cultural significance of nature tourism in the nineteenth-century United States, see John F. Sears, *Sacred Places: American Tourist Attractions in the Nineteenth Century* (New York: Oxford Univ. Press, 1989); on the cultural significance of American National Parks, see Alfred Runte, *National Parks: The American Experience*, 2nd ed. (Lincoln: Univ. Nebraska Press, 1987).

4. On expectations and the experience of the sublime, see David E. Nye, *American Technological Sublime* (Cambridge, MA: MIT Press, 1994), pp. 13–15; on amateur photography and notions of beautiful rural scenery in the early twentieth century, see John R. Stilgoe, "Popular Photography, Scenery Values, and Visual Assessment," *Landscape Journal*, 1984, 3(2):111–122. On the history of the National Parks, in particular nature preservation in the parks, see Richard West Sellars, *Preserving Nature in the National Parks* (New Haven: Yale Univ.

Press, 1997) and James A. Pritchard, *Preserving Yellowstone's Natural Conditions* (Lincoln: Univ. Nebraska Press, 1999).

5. Quote is from William T. Hornaday, *Popular Official Guide to the New York Zoological Park as Far as Completed* (New York: New York Zoological Society, 1900), p. v.

6. On American landscape painting in general, and on panoramas specifically, see Barbara Novak, *Nature and Culture: American Landscape and Painting, 1825–1875* (New York: Oxford Univ. Press, 1980).

7. F. Gruber, *Illustrated Guide and Catalogue of Woodward's Gardens* (San Francisco: Francis, Valentine & Co., 1880), pp. 64–65.

8. The narrative approach to natural history illustration developed alongside that of museum naturalists, whose illustrations showed animals out of context and emphasized diagrams showing animals' internal anatomy. On the genre division between narrative and systematic illustration, and on American natural history illustration in detail, see Ann Shelby Blum, *Picturing Nature: American Nineteenth-Century Zoological Illustration* (Princeton, NJ: Princeton Univ. Press, 1993). On Webster and Hornaday, on the translation of panorama painting to museum dioramas, and for a detailed history of museum taxidermy and display, see Karen Wonders, *Habitat Dioramas: Illusions of Wilderness in Museums of Natural History* (Uppsala: Acta Universitatus Uppsaliensis, 1993).

9. Hornaday, *Taxidermy and Zoological Collecting* (New York: Scribner's, 1891), p. 169.

10. Hornaday, *Taxidermy and Zoological Collecting*, p. 238; for biographical information on Hornaday, see, for example, James A. Dolph, "Bringing Wildlife to Millions: William Temple Hornaday, the Early Years, 1854–1896," Ph.D. Diss., University of Massachusetts, 1975, and Wonders, *Habitat Dioramas*; on the craft of taxidermy and the construction of dioramas, see Susan Leigh Star, "Craft vs. Commodity, Mess vs. Transcendence: How the Right Tool Became the Wrong One in the Case of Taxidermy and Natural History," in *The Right Tools for the Job*, ed. Adele E. Clarke and Joan H. Fujimura (Princeton, NJ: Princeton Univ. Press, 1992), pp. 257–286, and Donna Haraway, "Teddy Bear Patriarchy: Taxidermy in the Garden of Eden, New York City, 1908–36," in *Primate Visions: Gender, Race, and Nature in the World of Modern Science* (New York: Routledge, 1989), pp. 26–58.

11. *Washington Star*, 10 Mar. 1888, quoted in Dolph, *Bringing Wildlife to Millions*; on Hornday's bison group, see Wonders, *Habitat Dioramas*, pp. 120–122.

12. Haraway comments on the selective vision of wildlife—young, healthy, and often male—in habitat dioramas. See Donna Haraway, "Teddy Bear Patriarchy."

13. On the relationship between popular natural history and zoological illustration, see Ann Shelby Blum, *Picturing Nature*; on Seton, see Ernest Thomp-

son Seton, *Trail of an Artist-Naturalist: The Autobiography of Ernest Thompson Seton* (reprint, New York: Arno Press, 1978); Ernest Thompson Seton, "Communication Regarding the Needs of Artists in the Zoological Park," New York Zoological Society, *First Annual Report*, 1896:61–63.

14. On artists at the Bronx Zoo, see William Bridges, *A Gathering of Animals: An Unconventional History of the New York Zoological Society* (New York: Harper and Row, 1974), pp. 142–143.

15. Carl Rungius, "Seasonal Changes in the Form of the Rocky Mountain Sheep," *Zoological Society Bulletin*, 1913, 16(55):950–952; on Hornaday's relationship with Rungius, see Bridges, *A Gathering of Animals*, pp. 153–154.

16. Hornaday seemed not to consider that people might also want to take pictures of each other in the park, or of the buildings.

17. "Wild-Animal Photography," *Zoological Society Bulletin*, 1904, no. 12:133.

18. "Wild-Animal Photography," *Zoological Society Bulletin*, 1905, no. 16:194–195.

19. See, for example, the photo album "Views in the New York Zoological Park," 1901, and the souvenir book, "Decennial of the New York Zoological Park," 1909, NYZS, The Wildlife Conservation Society Archives. Presenting animals to the public in nature films required similar editing; see Gregg Mitman, *Reel Nature: America's Romance with Wildlife on Film* (Cambridge, MA: Harvard Univ. Press, 1999).

20. Monte Reinhart Hazlett, "Life Prisoners in Our Zoos," *Parks & Recreation*, 1918, 1(2):15.

21. Richard A. Addison, "Showmanship and the Zoo Business," *Parks & Recreation*, 1924, 8(2):129.

22. For the Hagenbeck quote, see Carl Hagenbeck, *Beasts and Men*, abridged translation by Hugh S.R. Elliot and A.G. Thacker (London: Longmans, Green, 1909), p. 40; for a detailed history of the Hagenbeck business, see Nigel T. Rothfels, *Savages and Beasts: The Birth of the Modern Zoo* (Baltimore, MD: Johns Hopkins Univ. Press, 2002); Rothfels discusses in detail the origins of Hagenbeck's panoramas and the ways they have been reinterpreted over time; see also David Ehrlinger, "The Hagenbeck Legacy," *International Zoo Yearbook*, 1990, 29:6–10.

23. The "fixed in the garden" quotation is from Hagenbeck, *Beasts and Men*, p. 41; on Hagenbeck's panoramas, see Hagenbeck, *Beasts and Men*; Rothfels, *Savages and Beasts*; Herman Reichenbach, "Carl Hagenbeck's Tierpark and Modern Zoological Gardens," *Journal of the Society for the Bibliography of Natural History*, 1980, 9(4):573–585.

24. On Sol Stephan's career, see David Ehrlinger, *The Cincinnati Zoo and Botanical Garden From Past to Present* (Cincinnati: Cincinnati Zoo and Botanical Garden, 1993), pp. 42–46; on John Benson, see Fletcher A. Reynolds, "Biographical Sketch of John T. Benson," *Parks & Recreation*, 1948, 31(11):655–656.

25. For Hornaday quote, see William T. Hornaday, "Observations on Zoological Park Foundations," *Parks & Recreation*, 1924, 8(2):124; on Hornaday's objections to Hagenbeck, see Bridges, *A Gathering of Animals*, p. 412.

26. A. Wetmore to W.M. Mann, 11 Feb. 1931, Records of the Office of the Secretary, RU 46, 1925–1949, box 144, folder 3, SIA.

27. Joseph A. Stephan, "Recent Developments in the Cincinnati Zoo," *Parks & Recreation*, 1939, 23(3):116.

28. Robert A. Bean, "Barless Enclosures: Progress of the Idea in the Zoos of the World," *Parks & Recreation*, 1925, 8(4):350.

29. Robert A. Bean, "Barless Enclosures," p. 350.

30. See Rothfels, *Savages and Beasts*, for a discussion of the ways Hagenbeck's panoramas have been reinterpreted over time.

31. Horace W. Peaslee, "Park Architecture: Zoological Gardens," *Architectural Record*, 1922, 51:365.

32. W.F.R. Mills, "Moving a Mountain to the City for Denver's New Habitat Zoo," *The American City*, 1918, 19(6):474.

33. Marguerite Mueller, "An Unappreciated Natural Rock Pile," *The St. Louis Zoo*, 1935, 6(2):3.

34. On Borcherdt, DeBoer, and the Denver Zoo in general, see Carolyn Etter and Don Etter, *The Denver Zoo: A Centennial History* (Denver: Denver Zoological Foundation, 1995); construction of the bear display is described in detail in Mills, "Moving a Mountain to the City."

35. On bear pits, see, for example, David Hancocks, *Animals and Architecture* (New York: Praeger, 1971), p. 115; quotations are from S.R. DeBoer, "Denver's New Zoo," *Parks & Recreation*, 1918, 2(1):3.

36. DeBoer, "Denver's New Zoo," pp. 3–7; see also Etter and Etter, *The Denver Zoo*, pp. 60–61.

37. DeBoer, "Denver's New Zoo," pp. 3–7; for a thoughtful exploration of the relationships among authenticity and nature, science, and popular culture, see Mitman, *Reel Nature*.

38. Mills, "Moving a Mountain to the City," p. 475; see also Horace W. Peaslee, "Park Architecture: Zoological Gardens."

39. On the expeditions, see Minutes of the Meetings of the Zoological Board of Control, 15 Oct. 1919, St. Louis Zoological Park Records, 1910–1941 (SL 194), Western Historical Manuscripts Collection, University of Missouri, St. Louis.

40. Frank Schwarz, "A Barless Zoo," *Parks & Recreation*, 1924, 8(2):138.

41. Henry C. Muskopf, "Barless Bear Dens: Bruin's Habitation in St. Louis a Triumph in Construction," *Parks & Recreation*, 1921, 5(1):55–58.

42. Quote is from an unidentified newspaper clipping; see also "First of the New Forest Park Bear Dens Dedicated," *St. Louis Post-Dispatch*, 14 June 1921; both are located in St. Louis Zoological Park Records, 1910–1941 (SL 194), volume 5 scrapbooks, Western Historical Manuscripts Collection, University of Missouri, St. Louis; on the zoo's fame, see "Zoo Spreads Fame of St. Louis

around the Globe, Director Says," *St. Louis Star*, undated clipping, SL 194, volume 6 scrapbooks, Western Historical Manuscripts Collection, University of Missouri, St. Louis.

43. Henry C. Muskopf, *Official Illustrations of the St. Louis Zoological Park* (St. Louis: Zoological Society of St. Louis, 1922), p. 41.

44. "Natural Rock Settings: St. Louis Zoo to Reproduce Natural Missouri Scenery," *Parks & Recreation*, 1921, 5(2):179.

45. On Cincinnati, see Joseph H. Stephan, "New Barless Grottoes at Cincinnati," *Parks & Recreation*, 1934, 18(3):111; on Evansville, see Gilmore M. Haynie, "For Barless Enclosures," *Parks & Recreation*, 1933, 16(5):239.

46. See Nye, *American Technological Sublime*, pp. 23,276,277.

47. Estimates of Appropriations, District of Columbia, 1929, in Records of the Office of the Secretary, 1892–1893, 1900–1947, Records Relating to Budget, RU 49, box 6, folder 5, SIA.

48. Richard A. Addison, "Establishing and Caring for an Exhibit of Reptiles," *Parks & Recreation*, 1926, 9(3):352.

49. "Snakes as Pets and Performers at Staten Island Zoo," *Parks & Recreation*, 1942, 25(10):411.

50. Addison, "Establishing and Caring for an Exhibit of Reptiles," p. 352.

51. On the relationship between outdoor recreation and snake bite increase, see Raymond L. Ditmars, "Occurrence and Habits of Our Poisonous Snakes," *Bulletin of the Antivenin Institute of America*, 1927, 1(1):4.

52. "Snakes as Pets and Performers at Staten Island Zoo," p. 412.

53. W. Reid Blair, *In the Zoo* (New York: Scribner's, 1929), p. 129.

54. Theodore E. Vance to N. Hollister, 18 Sept. 1923, RU 74, Records of the National Zoological Park, box 97, folder 11, SIA; Hollister replied, "As to the rattlesnake crossing a hair rope, I believe that, although it is a common belief that the snake will never do so, it has been many times proved that the rope does not interfere with the snake's movements in the least." Hollister to Vance, 10 Oct. 1923, RU 74, box 97, folder 11, SIA.

55. L.M. Klauber, "A Herpetological Review of the Hopi Snake Dance," *Bulletins of the Zoological Society of San Diego*, 1932, no. 9 (24 Jan): 2; for an analysis of the rattlesnake in American culture, see David Scofield Wilson, "The Rattlesnake," in *American Wildlife in Symbol and Story*, ed. Angus K. Gillespie and Jay Mechling (Knoxville: Univ. Tennessee Press, 1987), pp. 41–72.

56. For example, "Horned Toad Lives, too Tired to Care," *New York Evening Post*, 24 Feb. 1928.

57. On reptile house education programs, see "Snakes as Pets and Performers at Staten Island Zoo," pp. 411–414; "Portland Zoo Has Snake Class," *Parks & Recreation*, 1945, 28(5):280–281; for examples of the work of popular herpetologist-authors, see Raymond L. Ditmars, *The Reptile Book* (New York: Doubleday, 1907); *Reptiles of the World* (New York: Sturgis & Walton, 1910); Roger Conant, *A Field Guide to Reptiles and Amphibians of the United States and*

Canada East of the 100th Meridian (Boston: Houghton Mifflin, 1958); Carl Kauffeld, *Snakes and Snake Hunting* (Garden City, NY: Hanover House, 1957); *Snakes, the Keeper and the Kept* (New York: Doubleday, 1969); R. Marlin Perkins, *Animal Faces* (Buffalo: Foster & Stewart, 1944); Marlin Perkins and Peggy Tibma, *One Magic Night: A Story From the Zoo* (Chicago: H. Regnery Co., 1952); Marlin Perkins, *Zooparade* (Chicago: Rand McNally, 1954).

58. William M. Mann is quoted in a radio talk manuscript, 1930, RU 7293, Papers of William M. Mann and Lucile Quarry Mann, box 12, folder 9, SIA; the letter to the editor appeared in the *Washington Star*, 8 July 1929.

59. The quotation is from a letter from C.G. Abbot to W.M. Mann, 13 Apr. 1929, RU 46, box 144, folder 1, SIA.

60. Peter Guillery, *The Buildings of London Zoo* (London: Royal Commission on the Historical Monuments of England, 1993), pp. 12,15,16,35–37.

61. "New Reptile House at Capital Fulfills Zoo Chief's Dreams," *New York Herald Tribune*, 19 Mar. 1931.

62. "Society is Present at Opening of New House for Reptiles," *Washington Post*, 28 Feb. 1931; on Horsfall, see Wonders, *Habitat Dioramas*, p. 232.

63. Susan Leigh Star refers to dioramas as "not unlike devotional alcoves in cathedrals" in "Craft vs. Commodity," p. 278.

64. This comparison of the reptile house to Romanesque cathedrals is drawn from Heather P. Ewing, "An Architectural History of the National Zoological Park," Senior Essay in History of Art, Yale University, 1990; on "cathedrals of science," see Wonders, *Habitat Dioramas*, p. 10; on religious imagery in the Natural History Museum in London, which opened in 1881, and an analysis of its design and construction, see Carla Yanni, *Nature's Museums: Victorian Science & the Architecture of Display* (Baltimore: Johns Hopkins Univ. Press, 1999), pp. 111–146.

65. The Marlin Perkins quote is from Twelfth Annual Report of the St. Louis Zoological Park, 1929, St. Louis Zoological Park Records, 1910–1941 (SL 194), Western Historical Manuscripts Collection, University of Missouri, St. Louis; on the Bronx Zoo's reptile house, see Raymond Ditmars, "Artificial Snake Dens," *Zoological Society Bulletin*, 1912, no. 49:822–823; on Cincinnati, see *Architectural Record*, Nov. 1937, p. 29 and Ehrlinger, *The Cincinnati Zoo and Botanical Garden*, p. 69; on Baltimore, see Arthur R. Watson, "A Reptile House for Baltimore," *Parks & Recreation*, 1948, 31(10):596–598.

66. The quotation is from Jon C. Coe, "Design and Perception," p. 206; architect and zoo director David Hancocks takes zoos to task for creating cheap knock-offs of Hagenbeck designs in David Hancocks, *A Different Nature: The Paradoxical World of Zoos and Their Uncertain Future* (Berkeley and Los Angeles: Univ. California Press, 2001), pp. 67–73.

67. See Melissa Greene, "No Rms, Jungle Vu," *The Atlantic Monthly*, December 1987, pp. 62–78; Jon C. Coe, "Design and Perception: Making the Zoo Experience Real," *Zoo Biology*, 1985, 4:197–208.

68. Nye discusses this break in ordinary perception as essential to the experience of the sublime; the sublime is also "tinged with terror" in Nye's analysis. See Nye, *American Technological Sublime*, pp. 1–16; quotation is from John Coe, quoted in Melissa Greene, "No Rms, Jungle Vu," p. 62. Few people are taken in by the illusion, of course. "Has anyone really been immersed in a zoo exhibit and forgotten even momentarily they are in a zoo in the middle of the city?" asks Vicki Croke. "News reporters and photographers covering the opening of an exhibit appear to be the only ones ever fooled by it." But the ingenuity of the technique is as much an attraction as its "natural" look. The technological feat of maintaining a rainforest in Omaha, in February, for example, makes the achievement that much more admirable. See Vicki Croke, *The Modern Ark: The Story of Zoos, Past, Present, and Future* (New York: Scribner, 1997), p. 80; "hair stand up" quotation is from Jon C. Coe, "Design and Perception: Making the Zoo Experience Real."

69. On the heightened appeal to tourists of landscapes in danger of destruction, see Nye, *American Technological Sublime*, p. 287. Nye argues for a "consumer's sublime," an experience tourists seek at places like the Grand Canyon and national parks that increasingly "are appreciated not as signs of nature's immeasurable power and sublimity, but as contrasts to a civilization that threatens to overwhelm them."

Chapter Six
Zoos Old and New

1. See Melissa Greene, "No Rms, Jungle Vu," *The Atlantic Monthly*, December 1987, p. 77; Terry L. Maple and Erika F. Archibald, *Zoo Man: Inside the Zoo Revolution* (Atlanta, GA: Longstreet Press, 1993), pp. 1–2; Scott Allen, "Trouble at Franklin Park," *The Boston Globe Magazine*, April 28, 1996, pp. 14+; Francis Desiderio, "Raising the Bars: The Transformation of Atlanta's Zoo, 1889–2000," *Atlanta History* 2000, 43(4):32–35.

2. Terry Maple, "Toward a Responsible Zoo Agenda," in *Ethics on the Ark: Zoos, Animal Welfare, and Wildlife Conservation*, eds. Bryan G. Norton et al. (Washington, DC: Smithsonian Institution, 1995), p. 22; "animal slums" quote is from Desmond Morris, "Must We Have Zoos?" *Life*, 1968, 65(November 8): 78; for additional examples of the history of zoos as a story of progress in stages, see Vernon S. Kisling, Jr., "The Origin and Development of American Zoological Parks to 1899," in *New Worlds, New Animals: From Menagerie to Zoological Park in the Nineteenth Century*, eds. Robert J. Hoage and William A. Deiss (Baltimore, MD: Johns Hopkins Univ. Press, 1996), pp. 109–125; Vicki Croke, *The Modern Ark* (New York: Scribner, 1997; Linda Koebner, *Zoo Book: The Evolution of Wildlife Conservation Centers* (New York: Tom Doherty Associates, 1994).

3. "Dankly cheerless" quote is in "The New Zoos," *Newsweek*, June 1, 1970, p. 58.

4. On environmentalism in the United States, see Robert Gottlieb, *Forcing the Spring: The Transformation of the American Environmental Movement* (Washington, DC: Island Press, 1993).

5. For more details on institutional changes at zoos, see Jeffrey Nugent Hyson, "Urban Jungles: Zoos and American Society," Ph.D. Diss., Cornell University, 1999, pp. 415–473.

6. National Zoo director Theodore Reed sometimes escorted animals; see Fourth Oral History Interview with Theodore H. Reed, by Pamela Henson, transcript, 14 April 1989, p. 21, RU 9508, Smithsonian Institution Archives; see also "Forbidden Animals," *Parks & Recreation*, 1936, 20(1):51–52; "New Law on Importation of Animals," *Parks & Recreation*, 1948, 31(8):457–458; Richard Littell, *Endangered and Other Protected Species: Federal Law and Regulation* (Washington, DC: Bureau of National Affairs, 1992), pp. 7–14.

7. Ninth Oral History Interview with Theodore H. Reed, transcript, 13 October 1989, p. 51, RU 9508, SIA; Fifth Oral History Interview with Theodore H. Reed, transcript, 3 August 1989, p. 11, 9508, SIA RU.

8. On United States species-preservation legislation in the 1960s and 1970s, see Thomas R. Dunlap, *Saving America's Wildlife* (Princeton, NJ: Princeton Univ. Press, 1988), pp. 142–155.

9. Elizabeth Frank, "Introduction," unpublished manuscript, June 18, 2000.

10. On golden lion tamarins, see Ninth Oral History Interview with Theodore H. Reed, transcript, 13 October 1989, p. 55, RU 9508, SIA; "enormous and profitable traffic," in "The Shame of the Naked Cage," *Life*, 1968 65(November 8): 71; illegal trafficking in rare and exotic animals remains problematic: for a recent exposé, see Alan Green, *Animal Underworld: Inside America's Black Market for Rare and Exotic Species* (New York: Public Affairs, 1999).

11. The system, called ISIS, is described in U.S. Seal, D.G. Makey, D. Bridgwater, L. Simmons, and L. Murtfeldt, "ISIS: A Computerised Record System for the Management of Wild Animals in Captivity," *International Zoo Yearbook*, 1977, 17:68–70.

12. Frank J. Thompson, "Baby Bear Traits," *Forest and Stream*, 1879, 13(4):605,610.

13. "Memphis Breeding Success," *Parks & Recreation*, 1926, 9(3):365; Edmund Heller, "Polar Bears Reared in Milwaukee," in *Zoological Parks and Aquariums*, Vol. 1 (Rockford, IL: Parks & Recreation, 1932), p. 67; C. Emerson Brown, "Rearing Hippopotamuses in Captivity," *Parks & Recreation*, 1925, 8(3):280.

14. "Houston, Texas," *Parks & Recreation*, 1939, 22(11):589; Helen Martini, *My Zoo Family* (New York: Harper & Brothers, 1955); the women honored by the National Zoo were Margaret Grimmer, Louise Gallagher, Mrs. Herbert Stroman, Elizabeth Reed, Lucile Mann, and Esther Walker. See Sybil E. Hamlet, "The National Zoological Park from Its Beginnings to 1973," typescript, p. 262,

RU 365, Records of the National Zoological Park, Office of Public Affairs, box 37, folder 1, SIA. An exception to the rule against hiring women as paid keepers occurred, not surprisingly, during World War II, when Seattle's Woodland Park Zoo hired Margaret Wheeler and Melvina Kuempel. See "Seattle Has Women Zoo Keepers," *Parks & Recreation*, 1943, 26(8):385–387.

15. Erna Mohr, a German zoologist, discovered the potential for inbreeding by chance. A Hartmann's mountain zebra filly had been born in the Warsaw Zoo and named in her honor, and some years later Mohr decided to find out what had become of it, and also to investigate its pedigree. She found that the animal had been sold to a dealer, who sold it to another zoo, which bred the zebra with its father. "This is an extreme example of how any amount of care is still not sufficient to prevent inbreeding occurring unless the pedigree of individual animals can be ascertained *beforehand*," Mohr observed. Mohr was a fish systematist at the Hamburg Zoological Museum; she had an interest in zoos and their collections, however, and in 1925 began compiling a studbook for the wisent, the first studbook for an endangered species. On Mohr, see Chris Wemmer and Steven Thompson, "A Short History of Scientific Research in Zoological Gardens," in *The Ark Evolving: Zoos and Aquariums in Transition*, ed. Christen M. Wemmer (Front Royal, VA: Smithsonian Institution Conservation and Research Center, 1995), pp. 79–80; Erna Mohr, "Studbooks for Wild Animals in Captivity," *International Zoo Yearbook*, 1968, 8:159–167; on zookeeper beliefs, Elizabeth Frank, Curator of Mammals, Milwaukee County Zoo, personal communication; on the study at the National Zoo, see Croke, *The Modern Ark*, p. 162, and Jake Page, *Smithsonian's New Zoo* (Washington, D.C.: Smithsonian Institution, 1990), p. 129.

16. In 1994 the AAZPA changed its name to the American Zoo and Aquarium Association (AZA). Ullic Seal, at the Minneapolis U.S. Veteran's Hospital Metabolic Research Laboratory, deserves much of the credit for starting ISIS. See, for example, Seal et al., "ISIS: A Computerised Record System for the Management of Wild Animals in Captivity."

17. See Croke, *The Modern Ark*, p. 167; web site for the American Zoo and Aquarium Association (www.aza.org), April 2001.

18. On reproductive technology, see, for example, Croke, *The Modern Ark*; on cloning, see Jon Cohen, "Can Cloning Help Save Beleaguered Species?" *Science*, 1997, 276:1329–1330; quote is from Heini Hediger, in Terry L. Maple, "Toward a Unified Zoo Biology," *Zoo Biology* 1982, 1:1. "Zoo biology" is a term coined by Hediger, author of the "bible" of zoo management, *Wild Animals in Captivity* (New York: Dover, 1964). Captive breeding opens a minefield of ethical questions, for example, about how to care for zoo-bred populations of endangered animals whose habitats have been destroyed. A large literature discusses such issues; see, for example, Bryan G. Norton, Michael Hutchins, Elizabeth F. Stevens, and Terry L. Maple, eds. *Ethics on the Ark: Zoos, Animal*

Welfare, and Wildlife Conservation (Washington, DC: Smithsonian Institution Press, 1995) and Stephen St. C. Bostock, *Zoos and Animal Rights: The Ethics of Keeping Animals* (New York: Routledge, 1993).

19. The Philadelphia Zoo was by no means the first to necropsy animals that died; a century earlier necropsies had been done, and tuberculosis discovered, in the zoological collection in Paris. Richard Burkhardt, Jr., personal communication. See also Charles B. Penrose, "Foreward," in *Disease in Captive Wild Mammals and Birds*, by Herbert Fox (Philadelphia: J.B. Lippincott, 1923), p. 1; Zoological Society of Philadelphia, 35th Annual Report, 1907, pp. 12–13; Zoological Society of Philadelphia, 37th Annual Report, 1909, p. 13; Zoological Society of Philadelphia, "An Animal Garden in Fairmount Park," typescript, 1988, p. 17. The question of whether tuberculosis could be transmitted between humans and animals remained controversial; it was the subject of an International Tuberculosis Congress held in London in 1901.

20. On Philadelphia, see Roger Conant, *A Field Guide to the Life and Times of Roger Conant* (Toledo, OH: Toledo Zoological Society, 1997), p. 138; on the Bronx, see W. Reid Blair, *In the Zoo* (New York: Charles Scribner's Sons, 1929), p. 29.

21. Smithsonian Institution, *Annual Report*, 1950: 90–91.

22. Theodore H. Reed, who joined the staff of the National Zoo in 1955 and later became director, recalled that zoos contracted with animal dealers for a kind of exclusive display right—when a zoo purchased a rare animal, the dealer promised not to sell that species to other zoos for a period of three years. See Fifth Oral History Interview with Theodore H. Reed, transcript, 3 August 1989, p. 11, RU 9508, SIA.

23. See, for example, Mary Jane Moore, "Boss Animal Man at National Zoo Retires after a Fabulous Career," *Springfield (MA) Sunday Union and Republican*, March 19, 1944; clipping in RU 7293, Box 8, SIA.

24. Belle J. Benchley, *My Life in a Man-Made Jungle* (Boston: Little, Brown, and Co., 1940), pp. 223–224; see also Dr. Reuben Hilty, "Value of Veterinary Service in a Zoological Garden," in *Zoological Parks and Aquariums*, vol. 1 (Rockford, IL: Parks & Recreation, 1932), p. 67.

25. W. Reid Blair, *In the Zoo* (New York: Charles Scribner's Sons, 1929), pp. 182,189,191; tiger, see Sybil E. Hamlet, "The National Zoological Park from Its Beginnings to 1973," p. 243; see also Fourth Oral History Interview with Theodore H. Reed, p. 7, transcript, 14 April 1989, p. 21, RU 9508, SIA.

26. See "The Story of Cap-Chur Equipment" and Palmer Chemical & Equipment Co., Inc., "Cap-Chur Equipment," RU 365, Records of the National Zoological Park, Office of Public Affairs, box 12, folder 1, SIA. A newspaper reported on unforeseen uses of the Cap-Chur gun: "Surrealist artist Salvador Dali has used one to fire paint at canvasses to get an 'explosive' effect, and half-crazed rioting convicts in jails have been quietened by firing drugs into

them," in *The Star*, Johannesburg, South Africa, July 20, 1962; clipping reprinted in "Cap-Chur Equipment" above.

27. Fourth Oral History Interview with Theodore H. Reed, pp. 1–2, 8, transcript, 14 April 1989, p. 21, RU 9508, SIA. Reed had been appointed veterinarian at the National Zoo in 1955. As a veterinarian with experience treating exotic animals he was unusual—while practicing in Portland, Oregon, he had sometimes taken care of zoo animals there. He became director in 1958.

28. Jill Mellen, "Technology and Applied Animal Management at Disney's Animal Kingdom," presented at the MIT Media Laboratory, Z001 An Animal Odyssey: Symposium on Technology, Zoos of the Future, and Lessons for Toy Design, January 17, 2001.

29. Grant R. Jones, Jon C. Coe, and D.E. Paulson, "Woodland Park Zoo: Long-Range Plan Development Guidelines and Exhibit Scenarios," Jones and Jones, for Seattle Department of Parks and Recreation, 1976; Jon C. Coe, "Design and Perception: Making the Zoo Experience Real," *Zoo Biology* 4(1985):197–208; *Woodland Park Zoo Guide* (Seattle: Woodland Park Zoological Society, 1994), p. 9.

30. The first quotation is from Fairfield Osborn, "The Opening of the African Plains," *Bulletin of the New York Zoological Society*, 1941, 44 (3):69; Gregg Mitman analyzes the place of the African Plains exhibit in conservation education in *Reel Nature*, pp. 85–87; second quotation is from Fairfield Osborn, "and now . . . the next fifty years" *Animal Kingdom*, 1949, 52(6):32.

31. The philosophy of William Carr, founder of the Desert Museum, was developed during his earlier experience establishing the nature trails and Trailside Museum at Bear Mountain, New York. William H. Carr, *Pebbles in Your Shoes: The Story of How the Arizona-Sonora Desert Museum Began and Grew* (Tucson, Arizona: Arizona-Sonora Desert Museum, 1982).

32. On Milwaukee, see George Speidel, "Philosophy of the New Milwaukee County Zoo," *Parks & Recreation*, July 1957 p. 20; on San Diego, see Douglas G. Myers with Lynda Rutledge Stephenson, *Mister Zoo: The Life and Legacy of Dr. Charles Schroeder* (San Diego: The Zoological Society of San Diego, 1999); Desmond Morris called for a revolution at zoos in "Must We Have Zoos?"; William G. Conway, "How to Exhibit a Bullfrog: A Bedtime Story for Zoo Men," in *International Zoo Yearbook*, ed. Nicole Duplaix-Hall, vol. 13 (London: Zoological Society of London, 1973), pp. 221–226.

33. William G. Conway, "How to Exhibit a Bullfrog: A Bedtime Story for Zoo Men."

34. The textbook is Kenneth J. Polakowski, *Zoo Design: The Reality of Wild Illusions* (Ann Arbor, MI: University of Michigan School of Natural Resources, 1987); the quotation is from Polakowski, p. 165; an article that applauds the contributions of landscape architecture to zoo design is Anne Elizabeth Powell, "Gardens of Eden," *Landscape Architecture* (April 1997), pp. 78–87+.

35. On enrichment see David J. Shepherdson, Jill D. Mellen, and Michael Hutchins, eds., *Second Nature: Environmental Enrichment for Captive Animals* (Washington, DC: Smithsonian Institution Press, 1998); quotation is from *Second Nature*, p. 3; "animal welfare science" was the subject of Don Moore, "The Evolution of Zoos," presented at the MIT Media Laboratory, Z001 An Animal Odyssey: Symposium on Technology, Zoos of the Future, and Lessons for Toy Design, January 17, 2001; on the AZA Animal Welfare Committee, see Anne Baker, "A Moral Responsibility," *AZA Communiqué* (January 2001):27.

36. The AZA is quoted from the organization's web site, www.aza.org; John C. Coe, "Design and Perception: Making the Zoo Experience Real," *Zoo Biology*, 1985, 4:197; "on the defensive," is from Anne Baker, "A Moral Responsibility."

37. Leonard Woolf, *Downhill all the Way: An Autobiography of the Years 1919–1939* (London: The Hogarth Press, 1967), p. 42.

38. On humanitarians and animal welfare, see Lisa Mighetto, *Wild Animals and American Environmental Ethics* (Tucson: Univ. Arizona Press, 1991), pp. 53–74; James M. Jasper and Dorothy Nelkin, *The Animal Rights Crusade: The Growth of a Moral Protest* (New York: The Free Press, 1992), pp. 56–60.

39. See Mighetto, *Wild Animals and American Environmental Ethics*, pp. 71–73; John Wuichet and Bryan Norton elaborate on welfare and authenticity criteria for animal welfare in the late twentieth century. These two criteria also describe welfare arguments early in the century. See John Wuichet and Bryan Norton, "Differing Conceptions of Animal Welfare," in *Ethics on the Ark: Zoos, Animal Welfare, and Wildlife Conservation*, eds. Bryan G. Norton et al. (Washington, DC: Smithsonian Institution, 1995), pp. 239–242.

40. On the Jack London Club see Mighetto, *Wild Animals and American Environmental Ethics*, pp. 68–69; see also "The Jack London Club and the Theaters," *Our Dumb Animals*, June 1919, 52(1):4; Francis H. Rowley, "The Children's Elephants," *Our Dumb Animals*, May 1914, 46(12):188.

41. Alexander Pope, "Animal Life in the Zoo: The Modern Way of Keeping Wild Animals," *Scientific American* (April 24, 1915):390; the story about F.C. Selous is recounted in Edward Hindle, "Foreward," in Heini Hediger, *Wild Animals in Captivity*, trans. G. Sircom (London: Butterworths Scientific Publications, 1950), p. vii.

42. The quotation is from Alexander Pope, "Animal Life in the Zoo: The Modern Way of Keeping Wild Animals," *Scientific American* (April 24, 1915):385; ideas about animal consciousness have a long history, which is beyond the scope of this chapter. For a popular discussion see Stephen Budiansky, *If a Lion Could Talk: Animal Intelligence and the Evolution of Consciousness* (New York: The Free Press, 1998).

43. "The Happy Prisoners," *Time* (September 18, 1950):72; Hediger's book provided a basis for rethinking the design of zoo exhibits from the perspective of animal behavioral needs. Zoo professionals regard it as a foundational text,

and the scientific journal *Zoo Biology* was founded as a forum for extending Hediger's ideas. Fairfield Osborn became president of the New York Zoological Society in 1940. He published *Our Plundered Planet* in 1948, in which he articulated his concerns about the environmental effects of human overpopulation. He was the son of Henry Fairfield Osborn, an early leader of the New York Zoological Society and an influential paleontologist at the American Museum of Natural History.

44. Fairfield Osborn, "The Good Life in the Zoo," *Reader's Digest* 82(February 1963):148; Max Eastman, "Don't Pity the Animals in the Zoo," *Reader's Digest* 58(March 1951):72.

45. Desmond Morris, "Must We Have Zoos?"

46. On paperback sales of *Animal Liberation*, see James M. Jasper and Dorothy Nelkin, *The Animal Rights Crusade*, p. 177; Tom Regan, "Are Zoos Morally Defensible?" in Bryan G. Norton, et al., eds., *Ethics on the Ark*, p. 45.

47. Paola Cavalieri and Peter Singer, eds., *The Great Ape Project: Equality Beyond Humanity* (New York: St. Martin's Griffin, 1993).

48. Dale Jamieson, "Against Zoos," in *In Defense of Animals*, ed. Peter Singer (New York: Harper & Row, 1985), pp. 108–117; Dale Jamieson, "Zoos Revisited," in *Ethics on the Ark*, ed. Norton et al., pp. 52–66; Randy Malamud, *Reading Zoos: Representations of Animals in Captivity* (New York: New York Univ. Press, 1998), pp. 3,5; on the authenticity criterion for animal rights, see John Wuichet and Bryan Norton, "Differing Conceptions of Animal Welfare," in *Ethics on the Ark*, eds. Norton et al., pp. 235–250; for a variety of perspectives on animal welfare, animal rights, and conservation in the context of zoos, see the essays in Norton et al., eds., *Ethics on the Ark*.

49. See Patricia Leigh Brown, "The Warp and Woof of Identity Politics," *The New York Times*, March 18, 2001, p. 4WK. On the human-animal relationship, see for example, James Serpell, *In the Company of Animals* (New York: Basil Blackwell, 1986); Jennifer Wolch and Jody Emel, eds., *Animal Geographies: Place, Politics, and Identity in the Nature-Culture Borderlands* (New York: Verso, 1998); Robert J. Hoage, ed. *Perceptions of Animals in American Culture* (Washington, DC: Smithsonian Institution, 1989); Kathleen Kete, *The Beast in the Boudoir: Pet-keeping in Nineteenth-Century Paris* (Berkeley: Univ. of California Press, 1994); Yi-Fu Tuan, *Dominance and Affection: The Making of Pets* (New Haven: Yale Univ. Press, 1984). The human-animal relationship also enters prominently in recent works of environmental history, such as Andrew C. Isenberg, *The Destruction of the Bison* (New York: Cambridge Univ. Press, 2000) and Gregg Mitman, *Reel Nature: America's Romance with Wildlife on Film* (Cambridge, MA: Harvard Univ. Press, 1999). Its importance to historical scholarship is evidenced by a recent two-year seminar on the subject held at the Shelby Cullom Davis Center for Historical Studies at Princeton University, and it enters into the idea of "biophilia" as proposed by E.O. Wilson. For a recent exploration of the biophilia

idea see Stephen R. Kellert, *Kinship to Mastery: Biophilia in Human Evolution and Development* (Washington, DC: Island Press, 1997).

50. Conway is quoted in Francis X. Clines, "What's 3 Letters and Zoologically Incorrect?" *The New York Times*, February 4, 1993, pp. A1,B4; "pompous," in "A Zoo by Any Other Name," *The New York Times*, February 5, 1993, p. A26; "wanna go," in Francis X. Clines, "Not a 'Zoo'? Only in New York," *The New York Times*, February 7, 1993, p. E2.

INDEX

Page numbers appearing in italics refer to illustrations.